Biologic Markers

in

Pulmonary Toxicology

Subcommittee on Pulmonary Toxicology
Committee on Biologic Markers

Board on Environmental Studies and Toxicology

Commission on Life Sciences

National Research Council

NATIONAL ACADEMY PRESS
Washington, D.C. 1989

NATIONAL ACADEMY PRESS   2101 Constitution Avenue, NW   Washington, DC 20418

NOTICE: The project that is the subject of this report was approved by the Governing Board of the National Research Council, whose members are drawn from the councils of the National Academy of Sciences, the National Academy of Engineering, and the Institute of Medicine. The members of the committee responsible for the report were chosen for their special competences and with regard for appropriate balance.

This report has been reviewed by a group other than the authors according to procedures approved by a Report Review Committee consisting of members of the National Academy of Sciences, the National Academy of Engineering, and the Institute of Medicine.

The National Academy of Sciences is a private, nonprofit, self-perpetuating society of distinguished scholars engaged in scientific and engineering research, dedicated to the furtherance of science and technology and to their use for the general welfare. Upon the authority of the charter granted to it by the Congress in 1863, the Academy has a mandate that requires it to advise the federal government of scientific and technical matters. Dr. Frank Press is president of the National Academy of Sciences.

The National Academy of Engineering was established in 1964, under the charter of the National Academy of Sciences, as a parallel organization of outstanding engineers. It is autonomous in its administration and in the selection of its members, sharing with the National Academy of Sciences the responsibility for advising the federal government. The National Academy of Engineering also sponsors engineering programs aimed at meeting national needs, encourages education and research, and recognizes the superior achievements of engineers. Dr. Robert M. White is president of the National Academy of Engineering.

The Institute of Medicine was established in 1970 by the National Academy of Sciences to secure the services of eminent members of appropriate professions in the examination of policy matters pertaining to the health of the public. The Institute acts under the responsibility given to the National Academy of Sciences by its congressional charter to be an adviser to the federal government and, upon its own initiative, to identify issues of medical care, research, and education. Dr. Samuel O. Thier is president of the Institute of Medicine.

The National Research Council was organized by the National Academy of Sciences in 1916 to associate the broad community of science and technology with the Academy's purposes of furthering knowledge and advising the federal government. Functioning in accordance with general policies determined by the Academy, the Council has become the principal operating agency of both the National Academy of Sciences and the National Academy of Engineering in providing services to the government, the public, and the scientific and engineering communities. The Council is administered jointly by both Academies and the Institute of Medicine. Dr. Frank Press and Dr. Robert M. White are chairman and vice chairman, respectively, of the National Research Council.

The project was supported by the Environmental Protection Agency; the National Institute of Environmental Health Sciences; and the Comprehensive Environmental Response, Compensation, and Liability Act Trust Fund through cooperative agreement with the Agency for Toxic Substances and Disease Registry, U.S. Public Health Service, Department of Health and Human Services.

Library of Congress Cataloging-in-Publication Data

National Research Council (U.S.). Subcommittee on
    Pulmonary Toxicology.
    Biologic markers in pulmonary toxicology.

    Includes bibliographical references.
    1. Pulmonary toxicology. 2. Biochemical markers-
Diagnostic use. I. Title. {DNLM: 1. Air Pollutants-
toxicity. 2. Biological Markers. 3. Respiratory
System-drug effects. 4. Respiratory System-
pathology. 5. Respiratory System-physiopathology.
WA 754 N295b]
RG732.N38 1989   616.2'0071   89-37280
ISBN 0-309-03992-4
ISBN 0-309-03990-8 (pbk.)

Subcommittee on Pulmonary Toxicology

Committee on Biologic Markers

Board on Environmental Studies and Toxicology

Commission on Life Sciences

Sponsors

National Heart, Lung, and Blood Institute
National Institute of Environmental Health Sciences
U.S. Environmental Protection Agency
U.S. Public Health Service, Agency for Toxic Substances and Disease Registry

Government Liaison Group

John R. Fowle III, *Chairman,* U.S. Environmental Protection Agency, Washington, D.C.
Henry Falk, Centers for Disease Control, Atlanta, Georgia
W. Harry Hannon, Centers for Disease Control, Atlanta, Georgia
Suzanne Hurd, National Heart, Lung and Blood Institute, Bethesda, Maryland
Dennis Jones, Agency for Toxic Substances and Disease Registry, Atlanta, Georgia
James Lamb, U.S. Environmental Protection Agency, Washington, D.C.
George Lucier, National Institute of Environmental Health Sciences, Research Triangle Park, North Carolina
Carol Mapes, Food and Drug Administration, Washington, D.C.
Michael D. Waters, U.S. Environmental Protection Agency, Washington, D.C.

Preface

The American people have become increasingly aware of the potential for exposure to toxic material in the air we breathe and have developed a need for accurate, objective information on the health effects of inhaled pollutants. In keeping with that need, the National Heart, Lung, and Blood Institute, the National Institute of Environmental Health Sciences, the Office of Health Research of the U.S. Environmental Protection Agency, and the Agency for Toxic Substances and Disease Registry of the U.S. Public Health Service asked the Board on Environmental Studies and Toxicology in the National Research Council's Commission on Life Sciences to examine the potential for use of biologic markers in environmental health research. The term "biologic markers" refers to indicators of events in biologic systems or samples. It is useful to classify biologic markers into three types—markers of exposure, of effect, and of susceptibility—and to describe the events peculiar to each type.

The Committee on Biologic Markers was organized and considered the subjects of environmental research in which the use of biologic markers had the greatest potential for major contributions. Three biologic systems were chosen: the reproductive system, the respiratory system, and the immunologic system. This report is the product of the Subcommittee on Pulmonary Toxicology.

The subcommittee comprised a wide variety of persons working in the field of respiratory health effects, including clinicians, epidemiologists, toxicologists, physiologists, pathologists, aerosol scientists, and biochemists. The intent was to consider various kinds of basic research wherein phenomena under study might have the potential for use as markers of environmental exposures and disease, even if the original goal of the research had nothing to do with such markers. The major topics of discussion at the first meeting of the subcommittee were the meaning of the term "biologic marker" in the respiratory system and how these markers could be used in environmental health research. Eventually, the subcommittee decided to place major emphasis on biologic markers of two types: markers of exposure and markers of effects of environmental pollutants. Markers of susceptibility to environmental materials were also considered important, but, because of limitations of resources, were included only if they could also serve as markers of exposure or effects.

Finding biologic markers that can be used to measure the extent of exposure to and effects of environmental pollutants is not easy. In exposed persons, a marker must be able to detect an extremely small increase in a disease due to environmental exposures against a large background of the same or similar diseases from other causes. Inevitably, then, the subcommittee expanded its consideration of markers to examples of exposures in industrial situations. That was considered to be valid because of the potential use of such markers or their modifications in occupational applications.

The subcommittee decided to organize its report according to types of biologic markers (markers of exposure and of effects), rather than according to specific pollutants, on the grounds that it was more important to discuss general approaches than to attempt to compile a list of pollutant-specific markers.

In the course of the subcommittee's deliberations, many additional scientists were called on to provide information. The subcommittee especially recognizes the contributions of John Catrava, Medical College of Georgia, Augusta, Georgia; Delores Graham, U.S. Environmental Protection Agency, Research Triangle Park, North Carolina; Anton Jetten, National Institute of Environmental Health Sciences, Research Triangle Park, North Carolina; Barbara Meyrick, Vanderbilt University, Nashville, Tennessee; and Barry Peterson, University of Texas, Tyler, Texas.

This report could not have been produced without the untiring efforts of the National Research Council staff. Beulah Bresler was the administrative secretary and Linda Poore was the bibliographer. Devra Davis, Alvin Lazen, and James Reisa provided encouragement and helpful insight. Norman Grossblatt edited the report. Finally, the subcommittee gratefully acknowledges the persistence, patience, and expertise of Richard Thomas, the project director, in bringing this report to its final form.

ROGENE F. HENDERSON, *Chairman*
Subcommittee on Pulmonary Toxicology

Contents

Tables and Figures

Executive Summary

In the course of a year, an adult breathes approximately 7 million liters of air. Industrial workers breathe about 20 liters/minute, and runners can breathe up to 80 liters/minute. As part of the body's autonomic processes, breathing fulfills the vital function of exchanging the gases of oxygen and carbon dioxide. Along with air, respiration can bring in common air pollutants, such as toxic pollutants, respirable particulate matter, and the "criteria" air pollutants—nitrogen dioxide, sulfur dioxide, ozone, and carbon monoxide.

Original interest in reducing the intake of these pollutants came directly from episodes of severe air pollution that were clearly linked with immediate increases in deaths related to respiratory disorder, such as those in the Meuse Valley, Belgium, in 1930 and in Donora, Pennsylvania, in 1948. By December 1952, the "great fog" of London had killed an estimated 4,000 residents, many of whom were elderly persons who already had heart and lung problems. Sobered by such episodes, the United States developed an aggressive program for reducing air pollutants.

Still, subtle differences in pulmonary performance can occur and chronic pulmonary problems afflict one of every five persons; exposure even of healthy persons to environmental pollutants, even at their current magnitudes, plays a role. Such problems include reductions in the maximal amount of air that can be expelled, increases in chest tightness, wheezing, and possibly even cancer. Young children, elderly persons, and those with chronic diseases are likely to be especially vulnerable to pollution; respiratory symptoms can occur in them when concentrations of air pollutants increase even slightly.

In light of these longstanding and continuing concerns and corresponding concerns for other elements of public health, the Board on Environmental Studies and Toxicology (BEST) of the National Research Council's Commission on Life Sciences undertook a major investigation of the use of biologic markers in health research. At the request of the Office of Health Research of the U.S. Environmental Protection Agency (EPA), the National Institute of Environmental Health Sciences, and the Agency for Toxic Substances and Disease Registry, the Committee on Biologic Markers was formed to clarify the concepts and definitions of biologic markers. After completion of its task, the committee organized two subcommittees: the Subcommittee on Reproductive and Neurodevelopmental Toxicology (with individual panels on male reproduction, female reproduction, pregnancy, and neurodevelopment),

1

and the Subcommittee on Pulmonary Toxicology, which developed this report.

This executive summary encapsulates each of the seven chapters in the body of the report and presents a summary of the major conclusions and recommendations. Preceding the summary of the chapters are introductory sections that present concepts, definitions, and selected applications of biologic markers.

CONCEPTS AND DEFINITIONS

The committee has defined, as described in Chapter 1, the following concepts related to biologic markers. Biologic markers in the context of environmental health are indicators of events in biologic systems or samples. It is useful to classify biologic markers into three types—markers of exposure, of effect, and of susceptibility—and to describe the events peculiar to each type. A biologic marker of *exposure* is an exogenous substance or its metabolite or the product of an interaction between a xenobiotic agent and some target molecule or cell that is measured in a compartment within an organism. A biologic marker of *effect* is a measurable biochemical, physiologic, or other alteration within an organism that, depending on magnitude, can be recognized as an established or potential health impairment or disease. A biologic marker of *susceptibility* is an indicator of an inherent or acquired limitation of an organism's ability to respond to the challenge of exposure to a specific xenobiotic substance. Biologic markers of susceptibility are discussed in this report only insofar as they also can serve as markers of exposure or effect.

Once exposure has occurred, a continuum of biologic events can be detected. These events may serve as markers of the initial exposure, administered dose (circulating peak or cumulative dose), biologically effective dose (dose at the site of toxic action, dose at the receptor site, or dose to target macromolecules), altered structure or function with no ensuing pathologic effect, or potential or actual health impairment. Even before exposure occurs, biologic differences among humans might cause some to be more suscep-tible to environmentally induced disease. Thus, biologic markers are tools that can be used to clarify the relationship, if any, between exposure to a xenobiotic compound and health impairment.

Markers of Exposure

Exposure is the sum of xenobiotic material presented to an organism, whereas dose is the amount of the xenobiotic compound that is actually absorbed into the organism.

Blood flow, capillary permeability, transport into an organ or tissue, the number of receptor sites, and route of administration (which determines the path of the parent material or its metabolites in the body) all can influence absorbed dose or biologically effective dose. An inhaled carcinogen that is retained in the lung might produce tumors in the lung; if the same material were ingested and eliminated via the kidney, renal tumors might be produced. If the parent material is responsible for the observed toxicity, the amount of metabolite that reaches the target might be of no importance. If metabolites are responsible, however, metabolism in the liver, another target organ, or elsewhere as a result of metabolic cooperation between several tissues is an important determinant of absorbed dose and biologically effective dose.

Markers of Effect

For present purposes, the effects of an exposure on an organism (the responses of an organism to an exposure) are considered in the context of the relationship of exposure to health impairment or to the probability of health impairment. An effect is defined as an actual health impairment or recognized disease, an early precursor of a disease process that indicates a potential for health impairment, or an event peripheral to any disease process, but correlated with one and thus predictive of development of health impairment.

A biologic marker of an effect or response, then, can be any change that is qualitatively or quantitatively predictive of health impairment or potential impairment associated with exposure. Biologic markers are also useful to identify endogenous components or system functions that are considered to signify normal health, e.g., blood glucose. It is important to recognize, however, that the concentration or presence of such markers represents points on a continuum. Therefore, the boundaries between health and disease can change as knowledge increases.

Markers of Susceptibility

Some biologic markers indicate individual or population differences that affect the biologically effective dose of or the response to environmental agents independently of an exposure under study. An intrinsic genetic or other characteristic or a pre-existing disease that results in an increase in the absorbed dose, the biologically effective dose, or the target-tissue response can be a marker of increased susceptibility. Such markers include inborn differences in metabolism, variations in immunoglobulin concentrations, low organ reserve capacity, or other identifiable genetically determined or environmentally induced variations in absorption, metabolism, and response to environmental agents. Other factors that can affect individual susceptibilities include nutritional status of the organism, the role of the target site in controlling overall body function, the condition of the target tissue (whether disease is or was present), and compensation by homeostatic mechanisms during and after exposure. The reserve capacity of an organ to recover from an insult at the time of exposure can play an important role in determining the extent of an impairment.

EXTRAPOLATION FROM ANIMALS TO HUMANS

Extrapolations of data from animals to humans should be based on the most sensitive animal species tested, barring clear evidence that that species is toxicologically distinct from humans. Recent-

ly, EPA issued guidelines for evaluating exposure studies and other guidelines concerning various toxic effects (e.g., reproductive mutagenicity). Those guidelines provide a means to estimate data quality and stipulate the types of data that are necessary to estimate exposure for assessment evaluations.

Laboratory animals and humans can differ in structure, physiology, and pharmacokinetics. Thus, data from animals must be used carefully in determining health risks in humans. Important similarities and differences in relation to particle deposition and clearance between laboratory animals and humans must be considered when animal data are used to model human diseases. It has been particularly difficult to model asthma and emphysema. Furthermore, animal models of effects of suspected human carcinogens—such as radon, cigarette smoke, and arsenic—have been difficult to develop.

The toxicity of some chemicals is mediated by activation or detoxification biotransformation reactions. Inasmuch as biotransformation differs among species, it is important to establish whether the routes and rates of human and animal metabolic pathways are similar.

Health risks often are associated with combinations of effects in humans. For example, cardiovascular disease in humans can encompass atherosclerosis and hypertension. Although swine provide the most suitable animal model for studying spontaneous atherosclerosis, young rats might be most appropriate for studying hypertension. Estimating human risks necessarily would entail some appropriate combination of the relevant animal test systems.

A common source of uncertainty in risk assessment is the dose-response relationship at low doses or for rare effects. It is often impractical to conduct animal studies of effects at low doses, because large numbers of animals are required to detect a low incidence of an effect. Demonstrable health effects in humans, given the limits of epidemiology, often are associated with high doses. Sensitive molecular markers being developed will permit study of the relationship between exposure to chemicals at low ambient concentrations and the formation of a molecular marker predictive of human risk. The development of biologic markers might enable scientists to make better use of laboratory-animal data in estimating the effects of chemicals in humans.

As a 1986 National Research Council study on drinking water and health observed, the timing of exposure and the dose-response patterns in animal studies have important implications for extrapolating the resulting data to humans.

MARKERS OF EXPOSURE

As the portal of entry for airborne pollutants, the respiratory tract should be an excellent site for detecting pollutant-specific markers. From the fine cilia and mucosa of the nasal passages, through the trachea, to the bronchi, bronchioles, and moist alveolar membranes, the respiratory anatomy may be considered a system for detecting and identifying a variety of markers. Investigators seek markers of general pulmonary function or toxicity that reflect such deviations from normal conditions as altered breathing patterns and airway constriction. Specific exposure markers can be devised by sampling and examining tissue in the nose and lung (e.g., with nasal or bronchoalveolar lavage) and noting concentrations of pollutants of interest, such as diesel-exhaust particles, chrysotile asbestos, wood dust, fiberglass, wollastonite, iron, silica, and volcanic ash.

This report assesses various respiratory phenomena as markers of exposure, disease, and in some cases predisposition to disease. The respiratory system can respond to inhaled toxicants in only a few ways. Pollutants, antigens, infection, exercise, cold, and psychologic factors can all alter breathing patterns and lead to airway constriction; and persistent alteration of lung structure can occur in persons with chronic obstructive pulmonary disease (such as chronic bronchitis or emphysema, fibrosis, granulomatous disease, or neoplasia). None of those responses can be associated with

a specific etiology; each can result from a variety of causative agents or conditions. Yet some markers could be chosen as peculiar to a specific disease state and could be quantitatively relatable to the degree or stages of the disease.

Chapter 2 describes several new approaches to the use of biologic markers for providing information on respiratory tract dosimetry. The development of biologic markers of exposure to xenobiotics offers much promise. New molecular biologic techniques permit the measurement of such molecular markers as adducts formed with macromolecules in the body; the techniques can be used to detect adducted material in blood, urine, and tissue samples and are sensitive enough for the measurement of adducts formed with DNA or protein in cells washed from the respiratory tract or collected in sputum.

Innovative procedures, such as magnetopneumography, allow estimation of the lung burdens of some types of particles. Refined histologic techniques have revealed the cellular sites of deposition of inhaled particles in the lung and thus created the potential for calculating the dose to critical cells. And techniques for analyzing markers are well advanced; e.g., new techniques allow analysis of exhaled air, sputum, nasal lavage fluid, and bronchoalveolar lavage fluid for chemical evidence of exposure to specific pollutants. The field of mathematical modeling has advanced to the point where models now include physiologic measurements, such as blood flow rates, ventilation rates, metabolic rates, and both blood-air and blood-tissue partition coefficients. The models have made it much easier to extrapolate data from animal pharmacokinetic studies to predicted deposition and distribution in humans.

To make optimal use of the new techniques, we must determine the relationship between markers of exposure and the characteristics of the exposures that generate them. The markers usually yield only yes-no answers; that is, a particular exposure did or did not occur. But we need to determine the kinetic relationships between formation and breakdown of markers, so that we can use mathematical models to answer the question, "Given this amount of this marker in this tissue, what exposures could have produced the marker?" In addition, there is a need to explore such readily available respiratory tract fluids as nasal fluids and sputum for chemical markers of exposure to specific pollutants.

Chapter 2 briefly summarizes methods for collecting clinical information. The standardized questionnaire on respiratory disease can play an important role in identifying markers of potential respiratory disease. Neither the clinical history of a patient nor information obtained from a population with a standardized questionnaire constitutes biologic material, but such information is important in identifying the potential for respiratory disease. For instance, severe respiratory illness before the age of 2 years implies a likelihood of respiratory illness in later childhood; and persistent wheezing in childhood predicts abnormal pulmonary function in adulthood. Since the early 1950s, efforts to develop standardized procedures for gathering clinical and epidemiologic data on pulmonary health status have been in place. One of the most successful questionnaires has been the one developed by the British Medical Research Council and adopted by the American Thoracic Society. This questionnaire is used widely in the United States and throughout the world.

Chapter 2 shows the importance of determining the mechanisms by which environmental pollutants induce lung disease. What are the sites of toxic actions? How much of a given pollutant is required at a given site to produce a given toxic response? Knowledge of the mechanisms by which toxicity occurs should provide the most pertinent information on potential early markers of exposure to environmental pollutants and initial stages of response to them.

applicable for use in intact humans and animals. These measurements range from commonly used clinical tests of respiratory function to currently developing assays of the integrity of tissue barriers. Most of the measurements described in this chapter are relatively noninvasive and do not require anesthesia, catheterization, or collection of tissue samples. Several are readily adaptable to large-scale epidemiologic studies; however, some that require extensive or highly sophisticated equipment are best suited to studies of small groups.

The physiologic markers of early biologic effects in intact organisms can be described in four general categories:

• Markers of changes in respiratory (gas-exchange) function of the lung. These are derived from measurements of ventilation and its control, lung mechanical properties, intrapulmonary gas distribution, and alveolar-capillary gas exchange. A wide variety of such measurements have been developed and are in common use.

• Markers of increased airway reactivity, both to specific environmental agents and to standardized physiologic provocation. These are usually indexes of respiratory function. They are distinct from markers in the first category, in that the focus is not on gas exchange.

• Markers of change in clearance of particles. Measurement is used to examine an important set of respiratory defense mechanisms, and alterations of the mechanisms at times provide an early indication of an adverse effect of inhaled environmental agents.

• Markers of increased permeability of the air-blood barrier and of increased uptake of potentially injurious materials in the lung. Increase in permeability is sometimes an early feature of lung injury due to inhaled particles.

MARKERS OF PHYSIOLOGIC EFFECTS IN INTACT ORGANISMS

Chapter 3 describes the role of physiologic measurements of the functional status of the respiratory tract, which are

MARKERS OF ALTERED STRUCTURE OR FUNCTION

Chapter 4 describes two of the simplest methods of assessing alterations in lung structure: examination with the naked eye and microscopic examination. Of course, access to relevant materials in vivo is problematic, requiring in most cases direct access to lung tissue during surgical procedures. Lung tissue at autopsy is very helpful in establishing the burden of inhaled particles. Separate autonomic regions can be sampled, and quantitative data obtained. Apart from gross examination of lung tissue, both light and electron microscopic examination can be useful. Light microscopy has the advantages of speed, accuracy, and low cost. It also has some limitations, in that the types of measurements of tissue compartments are restricted. Such measurements are more easily made with electron microscopy. However, electron microscopy requires expensive equipment and more highly trained personnel. Newer techniques using electron and x-ray microscopy can also identify elements such as silica, cobalt, and nickel in human lung.

The proliferation of epithelial cells, fibroblasts, and macrophages can be measured microscopically. Proliferation of those cells can serve as markers of lung damage. For example, oxidant gases and chrysotile asbestos fibers cause rapid incorporation of tritiated thymidine into bronchiolar and alveolar cells. Detection and analysis of inhaled particles are possible with microscopic techniques. The size and distribution of particles in pulmonary tissue can be assessed with current morphometric techniques.

MARKERS OF INFLAMMATORY AND IMMUNE RESPONSE

Chapter 5 discusses markers associated with the inflammatory and immune responses of the respiratory tract. Inflammation in the lungs starts with the macrophage, the resident cell in the lower respiratory tract. The macrophage response is usually followed rapidly by an influx of neutrophils, which can damage tissue directly. Neutrophils are, however, most effective for destroying bacteria, and their presence constitutes a primary response to injury in any part of the body. The presence and timing of the neutrophil influx are good markers of inflammatory response in the respiratory tract. Neutrophils are seen in nasal washings and in other biologic samples from the respiratory system after exposure.

Another inflammatory cell, which is less commonly seen, is the eosinophil. It is attracted to the lung by chemotactic factors released during an inflammatory response. Like neutrophils, eosinophils can release chemotactic factors for other cells. They can also release several potential irritants, including oxygen radicals and leukotrienes.

The influx of neutrophils and eosinophils is usually seen in the first 3-7 days of an inflammatory response. If the inflammatory response is sustained, it is usually accompanied by a specific immune response. The immune response is generally mediated by lymphocytes, which migrate into the inflamed area within days of the original insult.

Apart from cells in the inflamed area that serve as markers, proteins and other products of cells can be detected. Those products often appear in predictable patterns that allow one to estimate the duration and intensity of the inflammatory response. Increases in proteins and other products in biologic fluids, such as bronchoalveolar lavage fluid, can serve as sensitive markers of the inflammatory response.

The human respiratory tract contains a complex array of host defenses—anatomic barriers, mucociliary clearance, phagocytic cells, and various components of cellular and humoral immunity—that collectively cleanse inhaled air and inactivate infectious and other injurious agents that are inhaled. In particular, the mucosal lining of the small airways and alveolar airspaces contains many components of the immune system that are important in protecting the normal lung. However, some of the components also play an important role in immunologic lung disease.

Antigen-antibody complexes are the basis of immune response. The initial phase of the immune response usually begins with antigen processing by phagocytes, such as macrophages. That includes degradation of foreign substances and exposure of lymphocytes to antigens, which stimulate the production of antibody, sensitized cells, or both. Chapter 5 describes the mechanisms of the immune response and the different cells and factors involved in the immune response that can be used as markers.

MARKERS OF CELLULAR AND BIOCHEMICAL RESPONSE

Chapter 6 discusses current and developing cellular and biochemical techniques that provide a possible source of markers peculiar to the lung and upper respiratory tract. The most important of the new techniques are bronchoalveolar lavage and nasal lavage. The predominant cell in bronchoalveolar lavage fluids from healthy people is the macrophage. However, in an inflammatory state, neutrophils, eosinophils, and mast cells might be present. Macrophages play an important role in the regulation of the immune cellular response, acting as both promoters and suppressors of events during inflammation. Thus, bronchoalveolar lavage fluid is ideally suited to the study of cells from respiratory tract and their response to chemical insult in both laboratory animals and humans. For example, alveolar macrophages secrete interleukin-1; the role of interleukin-1 as a mediator of inflammation is uncertain, and bronchoalveolar lavage is providing important mechanistic information on it.

Biochemical markers of pulmonary disease processes are also present in lavage fluid. Lactate dehydrogenase, a cytoplasmic enzyme released from damaged or lysed cells, can be used as a marker of cytotoxicity. An increase in this enzyme activity can be used to distinguish between toxic events and physiologic responses. An increase in serum proteins in the lavage fluid serves as an indicator of increased permeability of the alveolar-capillary barrier. The activity of hydrolytic and proteolytic enzymes released into the epithelial lining fluid has been shown to be correlated with the toxicity of inhaled particles. Biochemical and cellular markers of inflammation are useful in animal

toxicity studies to rank the toxicity of a series of compounds and to study the mechanisms of development of late-occurring lung diseases, such as fibrosis and emphysema.

Testing for immune cellular response has often used whole animals or humans. The interaction of several components of the immune system can be tested, and body responses can be measured, so one need not rely on extrapolation from results obtained in vitro. The examination of individual cellular components is described in detail in Chapter 6.

Exposure to environmental toxicants can cause damage in individual cells at the level of DNA and other cellular components. New techniques are now available to measure molecular markers of damage to cell components. Molecular markers have proved useful in the detection and diagnosis of some infectious diseases and in sickle-cell anemia and alpha- and beta-thalassemia, and they are being developed for other diseases, such as Duchenne muscular dystrophy, cystic fibrosis, Lesch-Nyhan syndrome, phenylketonuria, antithrombin III deficiency, and alpha$_1$-antitrypsin deficiency. The development and use of molecular markers to identify cellular responses to environmental toxicants will be important in increasing understanding of the mechanisms involved in pulmonary disease.

RECOMMENDATIONS

The recommendations of the Subcommittee have been divided into four major groups: exposure dosimetry, physiologic measurements, structural and functional measurements, and cellular and biochemical measurements.

Exposure Dosimetry

More information is needed on factors that affect the dosimetry of inhaled toxicants at specific sites along the respiratory tract. Specifically needed is information on the following:

• Regional deposition of inhaled pollutants at various sites along the respiratory tract. Considerable information is available on regional deposition of inhaled particles larger than 0.1 μm in aerodynamic diameter, but relatively little is known about factors that govern the deposition of inhaled gases, vapors, and ultrafine particles (smaller than 0.1 μm in aerodynamic diameter). Specific factors that affect deposition, particularly airway structure, need to be assessed in both laboratory animals and humans.

• Pollutant effects on clearance of deposited material. Dosimetry at specific sites in the respiratory tract depends both on how much is deposited at the site and on how quickly it is removed. Interspecies studies of regional clearance are required.

• The capacity of tissues at the site of deposition to metabolize a pollutant to a more toxic or a less toxic form. The toxicity of an inhaled organic compound might be increased by metabolic activity at some sites of deposition.

• Effects of chemical and physical characteristics of pollutants on the site of sequestration and on the induction of injury in the respiratory tract.

Physiologic modeling of the pharmacokinetics of inhaled materials in animals and humans shows promise for allowing extrapolation of dosimetry data between species, sexes, and regimens. Extension of that approach to the active metabolites of inhaled compounds would greatly increase our understanding of the toxicity of inhaled materials. Physiologic modeling also requires information on deposition, clearance, and metabolism.

One region of the respiratory tract that has received little attention, but is readily accessible for sampling, is the nose. The analysis of nasal rinses or sputum to detect exposures to specific pollutants could be useful. It must be applied with appropriate knowledge of dosimetry differences between the nose and the rest of the respiratory tract when different toxicants are inhaled.

Macromolecular adduct formation offers a promising new method of measuring exposure to organic chemicals that are reactive or can be metabolized to reactive forms. Further research on the kinetics of adduct formation and clearance is required to determine the relationship between exposure history and the concentration of adducts in blood or tissue samples. Most research has been on formation of adducts with DNA and hemoglobin. Adducts with other macromolecules, particularly those with site specificity, should be explored as markers of dose.

Physiologic Measurements

Existing physiologic tools need to be extended and new tests need to be developed and evaluated to focus on specific sites of action of environmental pollutants and specific effects of given pollutants. That requires evaluation of pathophysiologic correlates assessed initially in animals and later in humans, both in controlled exposure settings and in population-based samples.

Further research is required on the role of increases in nonspecific airway reactivity in identifying persons susceptible to environmental agents and on the role of increases in airway reactivity in the natural history of chronic obstructive pulmonary disease (COPD). The role of transient changes in airway reactivity in response to specific environmental agents needs to be assessed in regard to risk of development of COPD.

Links among alterations in particle clearance, environmental exposures, and development of lung disease need to be studied further to determine the usefulness of clearance as a marker of susceptibility and response.

Markers of early endothelial changes that would identify persons likely to develop acute or chronic vascular injury are needed. Markers of endothelial dysfunction that demonstrate toxicant specificity should be sought. More information is needed on how endothelial barrier function is correlated with nonbarrier functions. Refinements in techniques are needed to render them applicable to the screening of large numbers of people for vascular function.

Immunologic, biochemical, cytologic, and structural markers identified as related to specific lung injury need to be correlated with physiologic measures of respiratory function, airway reactivity, particle clearance, and indexes of air-blood barriers. Understanding of those relationships could be important in developing methods for assessing risks associated with environmental exposures.

Structural and Functional Measurements

Animal studies are needed for increasing understanding of the pathologic sequelae of particle disposition at specific sites in the lung. Examples of some analytic methods that can be made highly site-specific and cell-specific are morphometry, immunocytochemistry, histochemistry, and in situ hybridization (i.e., formation of RNA-DNA hybrids).

Research is required on the specific cell kinetics of response to injury. Labeling indexes determined by autoradiography are not adequate for describing cell kinetics. New techniques, such as a combination of autoradiography with morphometry to measure cell pool sizes before and after injury, can make it possible to determine changes in the entire cell cycle during lung injury.

Three-dimensional reconstruction of cells and tissues could establish changes in intracellular organelles and cell-cell relationships that result from exposure and injury. Such techniques as computer-assisted tissue reconstructions, time-lapse photography, and high-voltage electron microscopy can be applied to obtain data on cell function and cell regulation.

Lavage fluids should be analyzed to determine whether exposure to particles or gases changes chemotactic activity. Alterations could be due to depletion or activation of pulmonary C5, oxidants, arachidonic acid metabolites, growth factors, and other chemotactic factors that might be important markers of response.

Cell- and organ-culture models should be developed for extrapolating animal data on histologic changes to humans.

Findings on early histologic markers of exposure and injury in animals are difficult to apply to humans, because they require invasive techniques. New ways to maintain and use human cells obtained by bronchoalveolar lavage, transbronchial lung biopsy, and tracheal exploration need to be developed.

Cellular and Biochemical Measurements

Bronchoalveolar lavage has proved useful for evaluating lung inflammation, but further research is required to determine its utility for assessing pollutant exposure. Interspecies studies are needed to determine relationships between changes in bronchoalveolar-lavage fluid constituents and site-specific and pollutant-specific injury. Where applicable, clinical studies should be used for confirmation of results.

Cell and mediator changes found in bronchoalveolar and nasal-lavage fluid need to be related to physiologic and pathologic changes to assess their utility as biologic markers.

Nasal lavage needs to be investigated as a means of evaluating pollutant exposure and as an epidemiologic tool. The characteristics of bronchoalveolar-lavage fluid, nasal-lavage fluid, and whole-lung specimens need to be correlated in humans and animal models.

Monoclonal antibodies and molecular genetic techniques need to be applied to characterize types and functions of cells of the respiratory tract. As those techniques are introduced, relationships between phenotypic changes, pollutant exposure, and cell function should be established.

It would be of value to identify markers of susceptibility. Changes in cells and mediators in lavage fluid should be examined as possible markers of susceptibility.

Markers of the early events leading to late-stage disease should be developed to serve as molecular probes of mechanisms of disease.

1

Introduction

Breathing fulfills the vital function of exchanging the gases of oxygen and carbon dioxide. With oxygen, breathing brings in common toxic air pollutants, such as nitrogen dioxide, sulfur dioxide, ozone, carbon monoxide, and particulate matter. Because of concern for the potential impacts of these and other pollutants on public health, the National Heart, Lung, and Blood Institute, the Office of Health Research of the Environmental Protection Agency (EPA), the National Institute of Environmental Health Sciences, and the Agency for Toxic Substances Disease Registry asked the Board on Environmental Studies and Toxicology (BEST) in the National Research Council's Commission on Life Sciences to conduct a study of the scientific basis, potential use, and current state of development and validation of biologic markers in the respiratory system. BEST organized the Committee on Biologic Markers to examine the use of biologic markers in environmental health research. Three specific biologic systems or fields of research were chosen for study: the reproductive system, the respiratory system, and the immune system. This is the report of the Subcommittee on Pulmonary Toxicology, which was charged to review potential biologic markers in the respiratory tract.

Biologic markers, broadly defined, are indicators of variation in cellular or biochemical components or processes, structure, or function that are measurable in biologic systems or samples. For most purposes in environmental health research, the reason for interest in biologic markers is a desire to identify the early stages of disease and to understand basic mechanisms of exposure and response in research and medical practice.

The growth of molecular biology and biochemical approaches to medicine has resulted in the rapid development of markers for understanding disease, predicting outcome, and directing treatment. Many diseases are now defined, not by overt signs and symptoms, but by the detection of biologic markers at the subcellular or molecular level. For example, liver and kidney diseases are often diagnosed by measuring enzymes in blood or proteins in urine; lead poisoning can be diagnosed on the basis of blood lead concentrations and such biologic changes as increases in heme biosynthesis components in red cells and urine; and many inborn errors of metabolism, such as phenylketonuria, are diagnosed on the basis of cell biochemical findings, rather than expressed dysfunctions. The identification, validation, and use of markers in medicine and biology depend fundamentally on increased understanding of mechanisms of action and the role of molecular and biochemical processes in cell biology.

It is important to recognize that markers represent signals on a continuum between health and disease and that their definitions might shift as knowledge of the fundamental processes of disease progression increases. That is, today's markers of exposure might become tomorrow's markers of early biologic effect. What are perceived at first to be early signals of risk could come to be considered health impairments themselves because the predictive relationship is so strong; i.e., an early signal could represent an effect at a stage in the progression at which it is difficult to prevent a disease. Thus, biologic markers can be valuable in the prevention, early detection, and early treatment of disease.

There is growing interest in the use of biologic markers to study the health effects of exposure to environmental toxicants in clinical medicine, epidemiology, toxicology, and related biomedical fields. Clinical medicine uses markers to allow earlier detection and treatment of disease; epidemiology uses markers as indicators of exposure, internal dose, or health effects; toxicology uses markers to help determine underlying mechanisms of diseases, develop better estimates of dose-response relationships, and improve the technical bases for assessing risks at lower levels of exposure.

TYPES OF BIOLOGIC MARKERS

The Biologic Markers Committee has defined the following concepts related to biologic markers. Biologic markers are indicators of events in biologic systems or samples. It is useful to classify biologic markers into three types—markers of exposure, of effect, and of susceptibility—and to describe the events peculiar to each type. A biologic marker of *exposure* is an exogenous substance or its metabolite or the product of an interaction between a xenobiotic agent and some target molecule or cell that is measured in a compartment within an organism. A biologic

marker of *effect* is a measurable biochemical, physiologic, or other alteration within an organism that, depending on magnitude, can be recognized as an established or potential health impairment or disease. A biologic marker of *susceptibility* is an indicator of an inherent or acquired limitation of an organism's ability to respond to the challenge of exposure to a specific xenobiotic substance. Biologic markers of susceptibility are discussed in this report only insofar as they can also serve as markers of exposure or effect.

Once exposure has occurred, a continuum of biologic events can be detected. These events may serve as markers of the initial exposure, administered dose (circulating peak or cumulative dose), biologically effective dose (dose at the site of toxic action, dose at the receptor site, or dose to target macromolecules), altered structure or function with no ensuing pathologic effect, or potential or actual disease. Even before exposure occurs, biologic differences among humans might cause some individuals to be more susceptible to environmentally induced disease. Thus, biologic markers are tools that can be used to clarify the relationship, if any, between exposure to a xenobiotic substance and disease.

Markers of Exposure

Exposure is the sum of xenobiotic material presented to an organism, whereas dose is the amount of the material that is actually absorbed into the organism or reaches a target tissue or organ.

Blood flow, capillary permeability, transport into an organ or tissue, the number of receptor sites, and route of administration (which determines the path of the parent material or its metabolites in the body) all can influence absorbed dose or biologically effective dose. An inhaled carcinogen might produce tumors in the lung; if the same material were ingested and eliminated via the kidney, renal tumors might be produced. If the parent material is responsible for the observed toxicity, the amount of metabolite that reaches the target might be of no importance. If metabolites are responsible, however, metabolism in the liver, in another target organ, or elsewhere as a result of metabolic cooperation between several tissues is an important determinant of absorbed dose and biologically effective dose.

Markers of Effect

For present purposes, the effects of an exposure on an organism (the responses of an organism to an exposure) are considered in the context of the relationship of exposure to disease or to the probability of disease. An effect is defined as a health impairment, a precursor that indicates a likelihood of health impairment, or an event peripheral to any disease process, but correlated with one and thus predictive of development of disease.

A biologic marker of an effect or response, then, can be any change that is qualitatively or quantitatively indicative of health impairment or potential impairment (disease process) associated with exposure. Biologic markers are also useful to identify endogenous components or system functions that are considered to signify normal health. It is important to recognize, however, that the magnitude of such a component or function represents points on a continuum. Therefore, the boundaries between health and disease can change as knowledge increases.

Markers of Susceptibility

Some biologic markers indicate individual or population differences that affect the biologically effective dose of or the response to environmental agents independently of the characteristics of a particular exposure. An intrinsic genetic or other characteristic or a pre-existing disease that results in an increase in the absorbed dose, the biologically effective dose, or the target-tissue response after an exposure can be a marker of increased susceptibility. Such markers include inborn differences in metabolism, variations in immunoglobulin concentrations, low organs reserve capacity, and other identifiable genetically determined or environmentally induced variations in absorption, metabolism, and response to environmental agents. Other factors that can affect individual susceptibilities include the nutritional status of the organism, the role of the target site in controlling overall body function, the condition of the target tissue (whether disease is or was present), and compensation by homeostatic mechanisms during and after exposure. The reserve capacity of an organ to recover from an insult at the time of exposure can play an important role in determining the extent of an impairment.

VALIDATION OF BIOLOGIC MARKERS

The usefulness of a biologic marker must be validated by establishing the existence of a relationship between an environmental exposure and the biologic change of interest. Two characteristics of assessment determine the validity of a marker: sensitivity and specificity. Sensitivity is the extent to which a biologic marker indicates that a particular characteristic is present when it is present. If sensitivity is high, the probability of obtaining false-negative results is low. Specificity is the extent to which a biologic marker indicates that a particular characteristic is absent when it is absent. If specificity is high, the probability of obtaining false-positive results is low.

It is desirable for a marker to be as specific and as sensitive as possible. Specificity is needed to be able to associate a biologic marker with exposure to a specific pollutant. Sensitivity is required because many environmental exposures are airborne pollutants at low concentrations. The ideal biologic marker of an environmental exposure would be pollutant-specific, available for analysis with noninvasive techniques, detectable in trace concentrations or at very low activities, inexpensive to detect, and quantitatively relatable to the degree of exposure. Very rarely will all those qualities be available in a biologic marker. Most markers discussed in this report lack at least one of the attributes. Furthermore, many, if not most, diseases can be due to multiple causative agents, only some of which are environmental. Nevertheless, biologic markers provide valuable information that can improve our ability to determine the extent of environmentally induced respiratory disease.

Before many markers can be used for large epidemiologic field studies, efforts must be directed toward both the miniaturization of laboratory techniques and the preservation and banking of appropriate specimens. Those efforts initially must be carried out by laboratory scientists developing new biologic markers and—because of the intensity of the developmental efforts—will usually be of only second-order interest. Delays in taking the new techniques from the laboratory and applying them to population-based field studies will be inevitable.

A useful approach to the validation of biologic changes as markers is to use experimental studies in animals and clinical (human) studies to develop a matrix of information that enables one to make estimates for humans (Table 1-1). For example, markers of acute effects of short-term, low-concentration exposures to a pollutant can be detected in both animals and humans. A comparison of this information with markers of chronic effects resulting from long-term exposure of animals to the same pollutant could lead to the development of markers that are more predictive of health effects in chronically exposed humans (McClellan, 1986).

Numerous instances of clinical research and animal toxicology studies have both used and provided validation of biologic markers. Analysis of bronchoalveolar-lavage fluid is used to detect markers of an inflammatory response in the respiratory tract (Hunninghake et al., 1979b; Henderson et al., 1985a; Utell et al., 1985; Reynolds, 1987). Those markers have proved useful in diagnosing, staging, and planning therapeutic approaches to pulmonary disease in humans (Reynolds, 1987); in screening for pulmonary toxicity of inhaled pollutants in animals (Henderson et al., 1985a); and in detecting human responses to inhaled pollutants in short-term clinical studies (Utell et al., 1985). From such studies has come information on potential biologic markers of both the magnitude and the respiratory effects of environmental exposure.

It is important to distinguish between physiologic responses that represent normal defense mechanisms and responses that are predictive of disease. An organism might have a continuum of responses to a noxious agent—from responses that lead to removal of the agent or repair of initial injury to responses that indicate that irreversible damage has occurred or is destined to occur. Markers are needed to indicate where a given response lies on the continuum.

USE OF BIOLOGIC MARKERS IN THE RESPIRATORY TRACT

The potential usefulness of markers applies to all organ systems and tissues. In this report, discussion is limited to use of markers in the respiratory tract. The respiratory tract, as a portal of

TABLE 1-1 Example of a Matrix for Determining the Validity of a Biologic Marker

Species	Nature of Exposure	External Exposure	Internal Dose	Health Effect
A	Acute	X	X	X
	Chronic	X	X	X
B	Acute	X	-	X
	Chronic	-	-	-
Human	Acute	X	X	X
	Chronic	?	?	?

X = Marker determined.
- = Marker not yet determined.
? = Not yet tested.

entry for airborne pollutants into the body (and as a route of exit of some materials), should be advantageous for detecting pollutant-specific markers. If tissues or cells of the respiratory tract react chemically with an inhaled pollutant, it might be possible to detect reaction products in the lumen of the respiratory tract or in cells washed from it. Gaseous pollutants or the volatile metabolites of pollutants might be detected in the exhaled breath of exposed people.

A disadvantage of attempting to assess toxicant-induced changes in the respiratory tract, with respect to potential biologic markers of effects of environmental exposures, is the lack of specificity of pulmonary responses in relation to etiologic agents. The lung can respond to inhaled toxic materials in only a few ways:

• Altered breathing patterns and airway constriction. Altered breathing patterns are receptor-mediated responses; airway constriction can result from the direct action of a variety of stimuli or be mediated through neurogenic reflexes. Besides pollutants, potential stimuli include such agents as various antigens (in sensitized persons), infection, exercise, cold, and psychogenic factors.

• Cell injury leading to inflammation. The inflammatory response is characterized by an influx of inflammatory cells and an increased permeability of the alveolar-capillary barrier, which might lead to edema. Such an inflammatory response can also be induced by immunologic responses in a sensitized lung or by infectious agents.

• Persistent alteration of lung structure, such as fibrosis, chronic obstructive pulmonary disease (such as chronic bronchitis or emphysema), granulomatous disease, or neoplasia. None of these conditions or responses has a specific etiology; each can be a response to a variety of causative agents or conditions.

Thus, although biologic markers, such as changes in respiratory function, can be used to detect some structural alterations, it is not readily possible to associate a single functional alteration with a single causative agent. In the same manner, it might be easy to detect an inflammatory response in the respiratory tract by analyzing bronchoalveolar-lavage fluid or nasal-lavage fluid, but without additional information it is not easy to associate inflammation with a particular environmental exposure.

The following chapters discuss potential biologic markers of environmentally induced pulmonary disease. Some of the markers represent new techniques made possible by recent technologic advances. Others represent new uses of technique or procedures that have been used in other fields of research.

STRUCTURE OF THE REPORT

Chapter 2 examines markers of exposure. The deposition and clearance of inhaled material is discussed. New methods for monitoring inhaled material are reviewed as well as clinical techniques for assessing exposure. Chapter 3 examines current methods for assessing respiratory function. In addition, methods for assessing airway hyperactivity and injury to alveolar and vascular tissues are described. Chapter 4 discusses methods for structural assessment of whole lung, airways, and parenchyma.

Chapter 5 describes the mechanisms of inflammatory and immune response in the respiratory system and the associated markers. Chapter 6 continues the discussion developed in Chapter 5 but focuses on the cellular and biochemical responses observed as the lung responds to chemical insult. Finally, Chapter 7 provides the recommendations of the Subcommittee on Pulmonary Toxicology.

2

Markers of Exposure

In all matters of toxicologic concern, one must relate the effect of a toxicant to the dose required to cause the effect. Our ability to determine dose accurately varies widely with the kind of study involved—from epidemiologic studies, in which dose estimation must often be based on limited information from area monitoring, to studies performed in vitro, in which the dose to a subcellular organelle can be measured exactly. One way to obtain an accurate measure of dose is to measure biologic markers of exposure. A biologic marker of exposure is a substance measured in a compartment within an organism—an exogenous substance or its metabolite or a product of an interaction between a xenobiotic agent and some target molecule or cell. This chapter examines biologic markers of exposure.

Exposure is the sum of xenobiotic material presented to an organism, whereas dose is the amount of the material that is absorbed into the organism (internal dose) or the amount of active material that reaches the site of toxic action (in the case of biologically effective dose). Thus, there is an important distinction between exposure and dose that should be considered in the assessment of human exposure to toxic substances.

In many animal toxicologic studies and human clinical studies, the internal dose is inferred from knowledge of the exposure to a toxicant administered by ingestion or injection. However, if the toxicant is volatile, much of the compound might be exhaled unchanged soon after ingestion or intraperitoneal injection, and the retained dose could be less than the administered dose. If the toxicant is delivered by inhalation, the internal dose is difficult to assess and will depend on exposure conditions (air concentration, timing of exposure, etc.), deposition and absorption efficiencies of the inhaled material, metabolism, and ventilation patterns of the exposed subject.

In inhalation exposures, the estimated dose is sometimes expressed as duration of exposure to a toxicant at a given atmospheric concentration. It is particularly important for regulatory agencies to be able to relate the appearance of an adverse health effect to specific atmospheric concentrations of a toxicant, because regulations are usually set in terms of allowable concentrations in air. In the absence of toxicokinetic data, however, atmospheric concentrations used in toxicologic studies can be misleading, if the internal dose is not linearly related to the atmospheric concentration, especially at the high concentrations often used in such studies.

If the mechanism of a toxic effect is

17

known (the ideal situation), one can speak of the biologically effective dose at the cellular or subcellular site of action; that is the actual dose causing the effect, but it might be difficult to measure, except in studies performed in vitro. Recent advances in molecular epidemiology (Perera and Weinstein, 1982; Wogan and Gorelick, 1985; Wogan, 1988) have provided techniques that allow sampling of such biologic markers as covalent adducts formed by the binding of toxicants with macromolecules at or near the site of toxic action. Such adducts could prove to be valuable markers of biologically effective dose (e.g., DNA and protamine adducts) or of exposure (e.g., hemoglobin adducts).

All those assessments of dose—i.e., biologically effective dose, dose to critical tissues, internal dose, and atmospheric concentration—are required for various aspects of health-effects studies, and biologic markers of each would be useful. Most important, however, is information on how each type of dose measurement is related to the others, so that measurement of some can be used to estimate others. Mathematical models based on the physical and chemical properties of the toxicant and the recipient organism have proved useful for extrapolation not only between different dose measurements, but also between species and between exposure regimens (NRC, 1987).

The mechanism of action of a toxicant is rarely known, and the dose to tissues where the toxic effects occur is usually the most useful expression of dose. Toxicokinetic studies are required to establish the relationship between an administered dose and the dose to the tissue of concern. It is useful if the relationship between the dose to a target tissue and the appearance of an indicator substance in readily sampled body fluids can be established, so that the dose to a target tissue can be estimated from available biologic material.

In the following sections, we first discuss the factors that govern the deposition of inhaled materials—particles, gases, and vapors; those are the factors that determine the initial internal dose resulting from inhalation exposure. We then discuss the toxicokinetics of the deposited material, i.e., the rate and extent of clearance of deposited material and its metabolites from the respiratory tract, of their distribution in the body, and of their excretion. We then treat methods and sites for monitoring for inhaled materials and their products in the body. All that information determines the dose to the critical tissue, or, if the mechanism of injury is known, can even reveal the biologically effective dose. Mathematical models used to extrapolate animal toxicokinetic data to humans are discussed, and the use of clinical techniques to assess markers of exposure and dose is reviewed at the end of this chapter.

DEPOSITION OF INHALED MATERIAL IN THE RESPIRATORY TRACT

Particles

Particulate matter can include solid, relatively insoluble particles and liquid droplets that can be readily soluble in body fluids. The deposition of both types of particles are governed by the same forces, but the disposition of the deposited material will depend on the chemical properties of the material. This section deals with the deposition of both types of particles. The clearance of insoluble particles and the clearance of lipid-soluble material, whether inhaled as an aerosol or as a gas or vapor, is dealt with later.

Inhaled particles can come into contact with airway surfaces and be deposited on them. The extent and site of deposition depend on various factors, such as particle characteristics, ventilation pattern, and airway structure. Deposition can occur by five basic physical mechanisms: impaction, sedimentation, Brownian diffusion, interception, and electrostatic precipitation.

• *Impaction* is inertial deposition. It occurs when a particle's momentum prevents it from changing course in an area where there is a change in the direction of airflow. Impaction is the main mechanism by which a particle having an aerodynamic equivalent diameter (D_{ae}) of 0.5 μm or more is deposited in the upper respiratory tract and at or near tracheobronchial-tree branching points. The probability of impaction is proportional to air velocity, rate of breathing, and particle density and size.

• *Sedimentation* is deposition due to gravity. When the gravitational force on an airborne particle is balanced by the total of forces due to air buoyancy and air resistance, the particle will fall out of the airstream at a constant rate, known as the terminal settling velocity. The probability of sedimentation is proportional to particle residence time in the airway and to particle size and density and is inversely proportional to breathing rate. Sedimentation is an important deposition mechanism for particles with D_{ae} of 0.5 μm or more, which penetrate to airways whose air velocity is relatively low, e.g., mid-size to small bronchi and bronchioles.

• Submicrometer-size particles, especially those with physical diameters of 0.2 μm or less, acquire a random motion due to bombardment by surrounding air molecules; this motion can result in contact with an airway wall. The displacement sustained by a particle is a function of the diffusion coefficient, which is inversely related to particle cross-sectional area. *Brownian diffusion* is a major deposition mechanism in airways whose airflow is low or absent, e.g., bronchioles and alveoli. However, extremely small particles can be deposited by diffusion in the upper respiratory tract, trachea, and larger bronchi.

• *Interception* is an important mechanism of deposition of fibers and occurs when a fiber edge makes contact with an airway wall. The probability of interception increases as airway diameter decreases; fibers that are long and thin can penetrate into distal airways before deposition. Fiber shape is also important, in that straight fibers penetrate more distally than do curved fibers.

• In *electrostatic precipitation*, some freshly generated particles are electrically charged and exhibit greater deposition than that expected from size alone. That is due either to ionic charges induced on the surface of an airway by the particles or to space-charge effects; repulsion of particles bearing like charges results in increased migration toward the airway wall. The effect of charge on deposition is inversely proportional to particle size and airflow rate. Most ambient particles become neutralized naturally because of the presence of air ions, so electrostatic deposition is generally a minor contributor to particle collection by the respiratory tract; however, it is important in some laboratory studies.

Patterns of deposition efficiency (i.e., percentage deposition of amount inhaled) in the human respiratory tract are shown in Figures 2-1 through 2-4. The use of different experimental methods and protocols results in considerable variation in reported values. Figure 2-1 shows the pattern for overall respiratory tract deposition. Note the deposition minimum over the size range 0.2-0.5 μm. As previously discussed, particles with diameters of 0.5 μm or more are subject to impaction and sedimentation, whereas the deposition of those 0.2 μm or less is diffusion-dominated. Particles with diameters between these values are minimally influenced by all three mechanisms and tend to have relatively long suspension times in air. They undergo minimal deposition after inhalation, and most are carried out of the respiratory tract in exhaled air.

The effect of breathing mode on particle deposition in humans is evident from Figure 2-1. Nasal inhalation results in greater total deposition of particles with diameters over 0.5 μm than does oral inhalation, because collection in the upper respiratory tract is greater. But there is little apparent difference in total deposition of particles of 0.2-0.5 μm between nasal and oral breathing.

Figure 2-2 shows the pattern of deposition in the upper respiratory tract (the larynx and airways above it). Again, it is evident that nasal inhalation results in much greater deposition than oral. The greater the deposition of a substance in

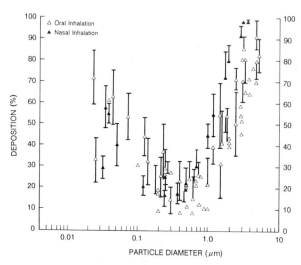

FIGURE 2-1 Particle deposition efficiency
in human respiratory tract. Values are
means with standard deviations. Source:
Reprinted with permission from Schlesinger, 1989.

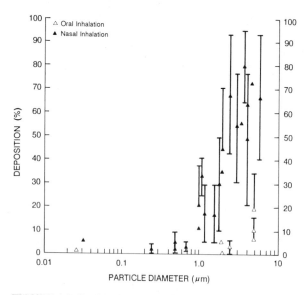

FIGURE 2-2 Particle deposition efficiency in
human upper respiratory tract. Values are means
with standard deviations. Source: Reprinted
with permission from Schlesinger, 1989.

the head, the less is available for removal in the lungs. Thus, the extent of collection in the upper respiratory tract affects deposition in more distal regions.

Figure 2-3 depicts deposition in the tracheobronchial tree. The relation between deposition and particle size is not as well defined as in other regions; fractional tracheobronchial deposition is relatively constant over a wide range of particle size.

Deposition in the pulmonary region (alveolated airways) is shown in Figure 2-4. With oral inhalation, deposition increases with particle size after a minimum at approximately 0.5 μm. With nasal breathing, percent deposition tends to decrease with increasing particle size. The removal of particles in more proximal airways determines the shape of the pulmonary deposition curves. For example, increased upper respiratory and tracheobronchial deposition would be associated with a reduction of pulmonary deposition; thus, nasal breathing results in less pulmonary penetration of larger particles, and a smaller fraction of deposition of entering particles, than does oral inhalation. In the latter case, the peak for pulmonary deposition shifts upward to larger particles and is more pronounced. However, with nasal breathing, there is a relatively constant pulmonary deposition over a wider range of particle size.

Rodents are often used in aerosol inhalation studies. To apply the results to humans adequately, it is essential to consider interspecies differences in total and regional deposition patterns. In evaluating studies with aerosols, the amount of deposition expressed merely as a percentage of the total inhaled—i.e., deposition efficiency—might not be adequate information for relating results between species. For example, total respiratory tract deposition for particles of the same size can be similar between humans and many laboratory animals; total deposition efficiency is independent of body size (McMahon et al., 1977; Brain and Mensah, 1983). Different species exposed to identical particles at the same exposure concentration will not receive the same initial mass deposition. If the total amount of deposition is divided by body weight, smaller animals would receive greater initial particle burdens per unit weight per unit exposure time than would larger ones.

Humans differ from most other mammals used in inhalation pharmacologic studies in various aspects of respiratory tract anatomy; but the implications for particle deposition have not been adequately understood. One major difference is in bronchial-tree branching pattern, which might affect the depth of penetration of inhaled particles, as well as localized patterns of deposition. In the pulmonary region, alveolar size differs between species; this could affect the probability of deposition by diffusion and sedimen-

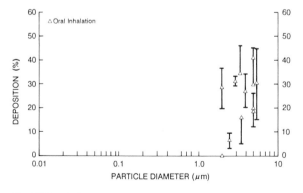

FIGURE 2-3 Particle deposition efficiency in human tracheobronchial tree. Values are means with standard deviations. Source: Reprinted with permission from Schlesinger, 1989.

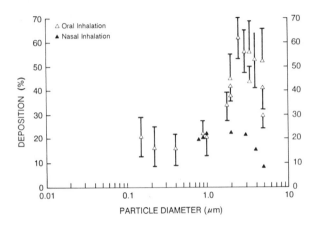

FIGURE 2-4 Particle deposition efficiency in human alveolated airways. Values are means with standard deviations. Source: Reprinted with permission from Schlesinger, 1989.

tation, because of interspecies differences in the distance between airborne particles and alveolar walls.

From the discussion of deposition mechanisms, it should be evident that the major particle characteristic that influences deposition is size. A particle characteristic that can alter its size after inhalation is hygroscopicity. Hygroscopic particles grow substantially while they are still airborne in the respiratory tract and are deposited according to their hydrated size, rather than their initial dry size.

Some environmental particles consist of a relatively insoluble core coated with various chemical substances, such as metals, acids, and organic compounds. Variations in both the core material and any adsorbed material depend on the source of the particles. For example, in combustion processes, volatile metal compounds and organic compounds might condense on carbon particles during the cooling of the effluent stream in the smokestack or exhaust line and during release to the atmosphere. Adsorption or condensation of gases from the atmosphere can produce a high surface concentration on particles that are already airborne. If those processes are diffusion-limited, the condensation and coagulation will be quantitatively proportional to particle diameter

for particles with D_{ae} larger than 0.5 μm and proportional to particle surface area for smaller particles. In either case, the fractional mass of the surface-coating material will be greater on smaller particles than on larger ones. Thus, surface deposition provides a layer of soluble material present at high concentration and results in small-particle enrichment, which leads to a shift in size of the potentially toxic surface materials to D_{ae} smaller than the D_{ae} of the total particle mass. Consequently, the deposition of surface-enriched material could be significantly increased.

Gases

Because of the rapid and consequent even distribution motion of gas molecules in air, inhaled gases can be deposited on or at least come into contact with a large portion of the surface of the respiratory tract. However, if deposited molecules are not removed by metabolic action, solubility in or chemical reaction with the epithelial lining fluid, or absorption into blood, the molecules will be re-entrained in the airflow and reach more distal parts of the respiratory tract. Deposition efficiency of a gas is usually defined as the proportion that is retained by the respiratory tract. Gases that are water-soluble and highly reactive chemically will be deposited (retained) to a higher degree in the upper respiratory tract, only at high concentrations will they reach the lung. Lipid-soluble, nonreactive gases, such as anesthetics, will be deposited in the alveoli, even at low concentrations.

The deposition and absorption of inhaled lipid-soluble gases have been investigated for some time by anesthesiologists (Fiserova-Bergerova, 1983). As noted above, lipid-soluble vapors are not absorbed by the aqueous lining fluid of the upper respiratory tract and so penetrate to the alveolar region at low concentrations. Their absorption from the respiratory tract depends mainly on their solubility in blood. Removal from the blood is by exhalation or metabolic action. A compound with low blood solubility and low tissue metabolic rate, such as vinylidene fluoride, will quickly reach saturation concentrations in the blood that depend on Henry's law and the gas-blood partition coefficient. The amount of such a compound that reaches other tissues depends on perfusion, or blood flow, because only a small amount will dissolve in blood. However, if a compound has high solubility in blood, such as acetone, the amount that reaches other tissues will depend on ventilation, because the blood will take up a large portion of any highly soluble chemical that enters the lung.

A ventilation-perfusion model similar to that developed for alveolar deposition of anesthetic (Henderson and Haggard, 1943, pp. 71-89) has been reported for the deposition of acetone and ethanol (nonreactive vapors) in the upper respiratory tract of the rat and guinea pig (Morris and Cavanagh, 1986; Morris et al., 1986). Species differences were observed: the rat was more efficient in deposition of the gases (Morris et al., 1986). Deposition of ethanol was twice as efficient as deposition of acetone, in agreement with the higher partition coefficient of ethanol and in agreement with the results of earlier studies of ethanol deposition in humans (Landahl and Hermann, 1950).

Water-soluble, reactive gases, such as the rat carcinogen formaldehyde (Kerns

et al., 1983) are deposited mainly in the nasopharyngeal region. Less water-soluble reactive gases, such as ozone and nitrogen dioxide, penetrate more deeply into the respiratory tract and damage the terminal bronchioles.

Miller and coworkers (F. J. Miller et al., 1982, 1985) have done extensive quantitative extrapolation modeling of the deposition of reactive gases in the respiratory tract. Their model takes into account not only the physical and chemical properties of an inhaled gas, but also the anatomic and physical properties of the respiratory tract and their influence on the deposition and absorption of inhaled gases.

The model predicts that dosimetry and uptake of ozone or other reactive gas will depend heavily on the thickness of the mucous blanket and on the rate of chemical reaction of the gas with the lining material. The model also predicts that exercise (with its increased ventilation rate) will increase the dose proportion of a reactive gas that reaches the lung.

CLEARANCE OF INHALED MATERIAL FROM THE RESPIRATORY TRACT

Insoluble Particles

The toxic response to inhaled particles depends on both the amount of material deposited at target sites and the duration of retention of deposited material. Particles are cleared from their deposition sites by various routes and interacting processes. The specific pathway depends on the region of the respiratory tract where the material is deposited.

The primary biologic mechanisms of clearance of insoluble particles are mucociliary transport in the nasal passages and tracheobronchial tree and removal from the pulmonary region by resident macrophages. Most of the surface of the tracheobronchial tree and nasal passages is lined with ciliated epithelium overlaid with mucus. The mucus is produced by specialized epithelial cells and submucosal glands and consists of two layers: a low-viscosity hypophase that surrounds the cilia and within which they move and a high-viscosity epiphase that lies on top of the cilia. Material deposited on the mucus is cleared to the pharynx by movement of the epiphase due to coordinated beating of the cilia.

Several mechanisms and pathways contribute to clearance from the pulmonary region, but their relative importance is uncertain. The mechanisms involve absorptive (dissolution) and nonabsorptive processes, which can occur simultaneously or at different times.

Nonabsorptive clearance processes are mediated primarily by alveolar macrophages. These large mononuclear cells originate as precursors in bone marrow, reach the lung as monocytes, and mature in the pulmonary interstitium, from which they traverse the epithelium to reach the alveolar surface. As macrophages move freely on alveolar surfaces, they phagocytose, transport, and detoxify deposited material, with which they come into contact by chance or by directed motion due to chemotactic factors.

Particle-laden macrophages are cleared from the pulmonary region along a number of pathways. The primary route is the mucociliary system, but the mechanism by which cells reach it is not certain. One possibility is movement along the alveolar epithelium; another involves passage through the alveolar epithelial wall into the interstitium—macrophages could then reach the surface of ciliated airways, perhaps through small collections of lymphatic tissue at alveolobronchiolar junctions.

Particle-laden macrophages that do not clear by way of the bronchial tree might actively migrate within the interstitium to a nearby lymphatic channel or, with uningested particles, be carried in the flow of interstitial fluid toward the lymphatic system. Alternatively, uningested particles or macrophages in the interstitium could cross the alveolar capillary endothelium and enter the blood directly. Finally, free particles or macrophages within the interstitium could end up in perivenous or subpleural sites, where they become trapped. The migration and grouping of particles and macrophages can lead to the redistribution of deposits into focal aggregates.

The most important mechanism of absorptive clearance is dissolution. Particles that dissolve in the alveolar fluid can diffuse through the epithelium and interstitium into the lymph or blood, and particles initially translocated to and trapped in interstitial sites can undergo dissolution there. Dissolution is a major clearance route even for particles usually considered to be relatively insoluble. The factors that affect the solubility of deposited particles are poorly understood, although they are influenced by the particles' surface-to-volume ratio and other surface properties. Some deposited material can dissolve after phagocytic uptake by macrophages. For example, some metals can dissolve within the acidic milieu of phagosomes. It is not certain, however, whether the dissolved material then leaves the cell.

The residence time of deposited particles depends on their clearance route. Material deposited on the conducting airways is cleared within about 1-2 days, although some long-term retention can occur. Particles deposited in the pulmonary region might remain for months or years or be retained indefinitely in interstitial sites. Soluble particles, and even particles with relatively low solubility, can dissolve in the pulmonary region. Solubilized components can be retained in the lungs, be redistributed in the body (where they might be retained in extrapulmonary tissues), or be excreted. In the conducting airways, solubilization occurs if the rate of dissolution is greater than the rate of removal by mucus transport. The mucociliary system of the lung provides a major line of defense in eliminating bacteria, inhaled particles, toxicants, and cellular debris. Bates (1989) has published an excellent review of the physiology of mucociliary clearance and its function in protecting the lung. The present subcommittee has addressed mucociliary clearance in several sections, depending on the type of contaminant and the type of marker being considered.

The retention of some materials cannot be studied experimentally in humans, so experimental animals must be used. Dosimetry depends on clearance rates and routes, so adequate pharmacologic assessment necessitates relating clearance kinetics in animals to those in humans. Although the basic mechanisms of respiratory tract clearance are similar in humans and most other mammals, regional clearance rates show substantial variation among species, even for similar particles deposited under comparable exposure conditions (Snipes et al., 1983). Dissolution rates and rates of transfer of dissolved substances into blood are probably related solely to the properties of the material being cleared and essentially independent of species (Cuddihy et al., 1979; Griffith et al., 1983; Bailey et al., 1985). However, different rates of mechanical transport, such as macrophage clearance from the pulmonary region (Bailey et al., 1985) and mucociliary transport in conducting airways (Felicetti et al., 1981), occur and result in species-dependent rate constants for these clearance pathways. Differences in regional (and perhaps total) clearance rates among species are probably due to the latter processes.

Gases and Soluble Particles

Reactive gases—such as ozone, nitrogen dioxide, and formaldehyde—exert their toxic effects at the site of deposition in the respiratory tract. Little is known about distribution of these gases or their products beyond the site of deposition. Therefore, the discussion of the clearance or disposition of inhaled materials in this section will include only nonreactive gases and soluble particles.

The collection of knowledge about the extent and rate at which inhaled toxic materials distribute throughout the body and are excreted is referred to as toxicokinetics. Toxicokinetic studies are fundamental to an understanding of internal dose and dose to target tissue. Toxicokinetic measurements are specific for individual pollutants and thus are valuable as biologic markers of environmental exposure.

For ethical and practical reasons, most detailed toxicokinetic studies have been performed in animals. In such studies, either radiolabeled material can be used to detect and measure the parent substance and its metabolites or standard analytic chemistry techniques can be used to measure the same materials. Using newer forms of mathematical modeling, which include physiologic parameters, one can make reasonable extrapolations from animal data to humans. This section discusses the types of toxicokinetic data that can be obtained in animal studies. The following sections discuss the types of human samples that can be analyzed to obtain information on exposure history, internal dose, and dose to target tissue and how animal toxicokinetic data can be extrapolated with modeling techniques to predictions for humans.

In animal inhalation exposure studies, one can determine the fraction of an inhaled substance that is absorbed, the time it takes to reach a steady-state concentration of the substance and its metabolites in the blood, equilibrium concentrations in tissues, major routes and rates of excretion of the substance and its metabolites, and times required for their elimination from each tissue and from the whole body. One can also determine the effects of exposure concentration, of exposure rate, and of repeated exposures on those measures. Tissue and excreta samples can be analyzed for materials of interest with standard analytic chemistry techniques or, for greater sensitivity, with radiolabeled compounds. In the latter case, the chemical form of a labeled compound (exposure material or metabolite) is often identified.

A few examples will illustrate the importance of such data in determining the internal dose of a compound received by an organism and the dose to target tissue. In rats exposed to methyl bromide at atmospheric concentrations of 50, 300, 5,700, and 10,400 nmol/L, the internal doses of the compound at the two highest exposure concentrations were equal (Medinsky et al., 1985). That was because the absorbance of methyl bromide and the tidal volume were decreased at the highest exposure concentration. The data indicate that absorption of methyl bromide is a saturable process and show the importance of knowing both internal dose and external exposure concentration.

When rats and mice were exposed by inhalation to formaldehyde at 14.3 ppm for up to 2 years, the rats had a 50% incidence of nasal carcinoma, the mice an incidence of only 1% (Kerns et al., 1983). The difference was explained biologically by analysis of effective dose, as opposed to administered dose, the external exposure concentration (Starr and Gibson, 1984). The mice were more sensitive to the sensory irritation properties of formaldehyde than the rats and thus had a smaller minute volume during exposure and received a lower internal dose (Barrow et al., 1983).

Such species differences can often be explained by toxicokinetic data, particularly if rates of formation and elimination of metabolites are determined. Studies in rats and mice exposed to benzene indicated that the mice had higher tissue and blood concentrations of putative toxic metabolites of benzene than did rats (Sabourin et al., 1987a,b). Mice were also more sensitive to benzene in long-term bioassay studies (NTP, 1986) and to the tumorigenic properties of inhaled butadiene (Huff et al., 1985). Pharmacokinetic studies on butadiene and its methylated derivative, isoprene, indicated higher blood concentrations of the reactive epoxide metabolites in mice (the more sensitive species) than in rats exposed at the same atmospheric concentration (Bond et al., 1986; Dahl et al., 1987).

It should be borne in mind that the lung is a target organ for some toxicants that can reach the lung through the blood or skin. For example, prolonged skin exposure, inhalation, or ingestion of the herbicide paraquat can cause death from lung injury in humans and animals. Paraquat accumulates selectively in lung tissue by a carrier-mediated mechanism and is retained there; accumulation not only influences lung paraquat burden, but is probably also an important determinant of organ response. Other pneumotoxic agents can reach the lung through the circulation, including antibiotics (e.g., bleomycin) and plant toxins (e.g., electrophilic metabolites of pyrrolizidine

alkaloids). Therefore, in the consideration of pulmonary markers and their development, it is important to examine not only inhaled environmental toxicants, but also those which reach the lung through the blood, through the skin, or by ingestion.

MONITORING FOR INHALED MATERIAL

Several biologic samples can be obtained from humans to assess internal dose or dose to target tissue. In reviewing the biologic approaches to dosimetry of carcinogens in humans, Tannenbaum and Skipper (1984) listed blood, urine, feces, sweat, hair, nails, milk, semen, saliva, lens, and biopsy tissues. Respiratory system samples—such as exhaled air, nasal-lavage fluid, and, in special cases, bronchoalveolar-lavage fluid— could be added. Substances most suitable for field sampling in epidemiologic studies are blood, urine, hair, nails, saliva, exhaled air, and perhaps nasal-lavage fluid. The other substances are more likely to be sampled in laboratory studies.

Biologic monitoring of industrial workers is most often based on blood, urine, or exhaled air (Lauwreys, 1983). Such an approach is appropriate in an industrial setting, because samples can be taken often and the exposures are normally higher than in an environmental setting. The methods also provide information for those in environmental research on the relationship between magnitude of exposure and the amount of compound or metabolite expected to appear in body fluids. However, for environmental exposures, such analyses might not be sensitive enough to detect small exposures; for compounds cleared rapidly from the body, only the most recent exposures can be detected. Some newer methods, however, have proved useful in monitoring for chemical exposures.

The following sections discuss the types of monitoring that can be done with such samples from humans and the kinds of animal studies on which some human monitoring is based. The emphasis is on newer techniques and on samples that are most relevant to exposure by environmental inhalation.

Insoluble Particles

The best marker of exposure to particles that one could hope for is detection of the inhaled material at the sites in the lung where disease develops. Establishing the presence of the particles at such sites is good; quantitation is better, if it is important to determine the dose delivered to a target site. In rats and mice exposed to asbestos, fiberglass, wollastonite, iron, silica, and ash from the volcano Mt. St. Helens, it has been established that 80% of the particles small enough to pass through the conducting airways was deposited on the bifurcations of alveolar ducts (Brody and Roe, 1983). Scanning electron microscopy was used after brief exposures (1, 3, or 5 hours), to calculate the number of particles per square micrometer of bifurcation surface (Brody and Roe, 1983). That number is a marker of exposure. Whether such a marker could be useful for human exposures is difficult to know. The lung would have to be fixed for electron microscopy within several hours after exposure; otherwise, substantial numbers of inhaled particles would have been transported from the alveolar surfaces by epithelial cells, macrophages, and the alveolar lining layer (Brody et al., 1981).

Particle deposition is a good marker of exposure in animals, because it can predict whether the subject is likely to develop lung disease, where the disease will originate, and the nature of the pathogenic process. The first prediction is based on the elemental nature of the inhaled particles as they reside on the epithelial surfaces. That is determined routinely with x-ray spectrometry, a technique widely used in studies of particle burden in humans and animals (Brody, 1984). If the particles are asbestos fibers or silica crystals, it would be valid to assume that a fibrogenic disease will ensue. If wollastonite fibers or ash particles are detected, it is less likely that a pathologic response will follow. The second prediction is validated by a series of studies that show that the initial response of epithelial cells, macrophages, and fibroblasts takes place at the sites of original particle deposition, i.e., the alveolar duct bifurcations (Warheit et al., 1986). Decades ago, pathologists used the finding of early inflammation and fibrogenesis in the bronchiolar-alveolar regions as a marker of particle-induced lung disease (Wagner, 1965). The nature of the response should be predictable on the basis of the two other predictions. If asbestos is inhaled, then interstitial macrophage-mediated fibrogenesis should be expected. If silica is inhaled, a nodular fibrosis mediated by acute and chronic inflammatory cells will develop. If iron or ash particles reach the alveolar surfaces, one should expect rapid clearance of particles with little or no pathogenic sequelae.

Some particles are retained in the lungs for long periods, even through the lifetime of the experimental animal or occupationally exposed person (Abraham, 1978). Such particles are excellent markers of exposure and, as suggested above, could increase understanding of the nature of any disease process that is present. For example, if rats inhale chrysotile or crocidolite asbestos for 1 hour, approximately 20% of it will still be in the lungs a month after exposure (Roggli and Brody, 1984; Roggli et al., 1987). If clearance continues as predicted by calculated clearance curves (Lippman et al., 1980; Roggli and Brody, 1984), the animals will still have many fibers in their lungs at the time of expected natural death. Occupationally exposed people have large quantities of dust in their lungs many decades after cessation of exposure (Selikoff and Hammond, 1978); this knowledge has yet to be exploited in a quantitative way in attempts to use lung burden to predict lung injury.

Many techniques are available for assessing the lung burden of many commonly inhaled particles. Several new imaging techniques can be used to determine the nature of crystalline particles. Transmission electron microscopy (TEM) is used to locate the particles in lung tissue; then energy-dispersive x-ray analysis (EDXA) and selected area electron diffraction (SAED) are used to identify the mineral species. Some techniques are noninvasive or minimally invasive and can be applied to humans, although definitive analysis of specific sites of particle burden requires access to properly fixed tissue.

• *Minimally invasive techniques for analysis of particle burden.* Use of labels, such as technetium-99, permits determination of particle size and total lung burden of inhaled particles with whole-lung scanning. In addition, single photon emission computed tomography (SPECT) scanning has sufficient resolution for routine determination of total particle burden in lung segments smaller than 1 cm^3 and can permit some definition of distribution of inhaled particles across the lung.

• *Bronchoalveolar lavage.* Numbers of fibers recovered from filtered aliquots of bronchoalveolar-lavage fluid have been shown to correlate closely with total lung asbestos burden (Roggli et al., 1986). Bronchoalveolar lavage might provide useful data on the lung burden of other types of particles, but the correlations have not been experimentally verified.

• *Open-lung biopsy.* Acid digestion of open-lung biopsy specimens permits direct counting of particles or determination of fiber burden per gram of lung tissue in selected patients. The obvious limitation of this marker is the need to justify an open-lung biopsy in a patient.

• *Lung ashing and assay for lung* particle burden. This technique permits a rigorous determination of total lung burden of some particles, such as asbestos fibers, but is applicable only to animal studies or to human autopsy studies.

• *Magnetopneumography.* This is a non-invasive technique by which magnetizable particles in the lung can be quantitated by the magnetic field induced by the particles. Because it is noninvasive, it is discussed in some detail.

Various inhaled particles contain minerals that can act as miniature permanent magnets, such as arc-welding fumes, foun-

dry dust, steel-mill dust, asbestos, fly ash, and coal-mine dust. In some cases—for example, the latter three—magnetic materials are minor constituents that can be used as tracers of the dust. The presence of magnetic particles in the thorax can be detected with an external measurement technique known as magnetopneumography, which involves the brief application of a strong magnetic field across the thorax to align the magnetic dipoles in the particles (i.e., to produce magnetization) and then the rotation of the magnetized particles into a common alignment. The resulting magnetic field, known as a remanent field, can then be measured with a magnetometer. The strength of the remanent field when the magnetizing field is turned off can be directly related to the amount of magnetic dust in the field of view of the magnetometer, provided that the spatial distribution of the dipoles in relation to the detector and the relation of magnetic moment to dust mass are known or can be estimated. The measurement of a magnetic field due to the presence of particulate contaminants in the human lung was first reported by Cohen (1973), and the verification that MPG can measure the retention of particles in the thorax was provided by later studies in experimental animals (Oberdorster and Freedman, 1981; Halpern et al., 1981).

A variety of techniques for initial magnetization and field-strength measurement exist; they differ in sensitivity, practicality for general or field use, and the influence of variations in the spatial distribution of dust in the thorax. All the techniques are characterized by uncertainty concerning the exact relationship between remanent field and lung burden; that is because the remanent field depends on both the magnetic moments of retained particles and their position in relation to the detector. For particles of mixed composition, the fraction that is magnetic usually varies with particle size, as do deposition and retention. The magnetic moment per unit mass that is assumed, or measured, for bulk dust samples probably differs from that of the fraction in the field of view of the detector, so there is a calibration error in estimates of burden.

Further errors result from differences between the assumed and actual distributions of particles in a field of view that includes areas of varied sensitivity.

There is always some time between magnetization and the measurement of remanent field. Once the external field is removed, the remanent field immediately begins to decay, as processes in the particle and the thorax act to randomize the orientation of the dipoles and particles; this loss of magnetic field with time is known as relaxation. Because relaxation begins immediately after removal of the field, the measured remanent-field strength is less than the initial, maximal value. Accordingly, the latter is determined by making serial measurements of remanent field and then extrapolating back to time zero. But even the extrapolated value might not indicate the total amount of dust in the field of view. For example, the presence of particles with relaxation times shorter than the time between the removal of the magnetizing field and the initial remanent-field measurement might not be detected. They could include small particles or particles in pulmonary macrophages that are being rapidly reoriented by intracellular activity. In addition, overall thoracic relaxation rates can vary widely because of a variable distribution of deposited dust in different lung compartments with time after inhalation.

The fraction of magnetic dust whose detection is prevented by extrapolating back can be expected to vary with time since deposition and with the amount of dust in the lungs. With increasing residence time, an increasing fraction of retained dust is in intracellular sites or in lymph nodes, where rotational freedom is diminished. With increasing dust burden, there is also an increased probability of aggregation of particles into clusters, which leads to diminution of rotational freedom.

In all MPG systems, the subject measured is moved relative to magnetic sensors. That is necessary, because strong, low-frequency variations of the local field often make continuous measurement of the remanent chest field impossible. In addition, the remanent field from dust in the lungs is much weaker than the earth's steady magnetic field or field variations due to environmental sources; thus, measurements must be performed in a magnetically shielded chamber or with a gradiometer field sensor that reduces sensitivity to both uniform background fields and local fluctuating fields. Table 2-1 shows some representative field strengths due to contaminants in the lungs, compared with background fields of the earth and the chest.

TABLE 2-1 Relative Strength of Various Magnetic Fields[a]

Source	Field Intensity, Picoteslas
Steady field of earth	70×10^6
Diamagnetic thoracic field	30
Urban dwellers	30
Asbestos workers	6.4-5,000[b]
Foundry workers	800[b]
Arc welders	50-3,000[b]
Machinists	27[b]
Coal miners	4.8-17[b]

[a] Compiled from Williamson and Kaufman (1981).
[b] Based on remanent fields in occupationally exposed persons.

Various measuring devices are used, and the amount of measurable material in the lungs depends on the specific sensor. Sensors range from large superconducting devices (superconducting quantum interference device, or SQUID) to smaller flux-gate magnetometers and induction coils (Table 2-2). Some of the systems allow mobility, which is useful in field studies (Kalliomaki et al., 1986). Induction coils show field sensitivity, which is related to field frequency; at some frequencies, they can be more sensitive than the SQUID magnetometer, but the spatial resolution is much poorer. Flux-gate magnetometers are relatively sensitive to field frequency and are ideal for inexpensively measuring fields due to large concentrations of magnetic particles in the lungs.

Although a simple system with one magnetometer and movement of the subject back and forth can be used, more sophisticated systems have increased the utility

TABLE 2-2 Sensitivities of Various Magnetic Detectors[a]

Detector	Sensitivity,[b] T/Hz
Induction coil	
Air core	5×10^{-12}
Ferrite core	3×10^{-13}
Flux-gate magnetometer	
Commercial	3×10^{-11}
Specialized	1.5×10^{-12}
Superconducting quantum interference device (SQUID)	
With shielded chamber	8×10^{-15}
Unshielded, urban area	8×10^{-14}

[a] Data from Williamson and Kaufman (1981).
[b] At 10 Hz.

of MPG. One such system (Kalliomaki et al., 1980, 1981) uses coaxial flux-gate gradiometers positioned both in front of and behind the subject, who rests on an inclined chair or platform and is moved horizontally between the detectors in increments to obtain a profile of remanent-field strength. A modification of that system that can determine the proportional burdens of different dusts has also been described (Junttila et al., 1983; Kalliomaki et al., 1983). Discrimination of the dusts is based on differences in "coercive" force (magnetic hardness), a characteristic of each specific metal. To determine coercive force, the thorax is magnetized to saturation with a strong field. A reversed magnetizing field is then applied, and the remanent field is measured. The strength of the reverse field required to just cancel the original remanent field is a measure of coercive force. The strength of the reverse field required for cancellation varies among different metals.

An approach that might allow greater precision in the measurement of thoracic dust burdens is localized field magneto-pneumography (LPG), whereby the field of view of the detector is matched to the size of the region magnetized (Robinson and Freedman, 1979). There is virtually no cancellation of the remanent field of the magnetic particles in the field of view by return lines from particles in adjacent tissues. After the region of interest is measured, the particles in

it are demagnetized, so that they do not interfere with later measurements in an adjacent region. LPG can be extended to provide information on the depth distribution of dust in the thorax by serial measurements of remanent field after successive degrees of demagnetization from the chest surface inward. Such information could be particularly important for long-term dust burdens, which can be concentrated in pleural and hilar lymph nodes. When such redistributions occur, thoracic measurements based on a calibration that assumes uniform distribution can show substantial errors.

Regardless of the specific MPG technique used, the external field must be strong enough and be applied long enough to magnetize the bulk of the deposited particles. Paired or single electromagnetic coils are generally used. In the uniform-field MPG, the whole thorax is magnetized. In LPG, one region is magnetized at a time, and the resulting field is measured and then erased before application of the magnetic field to the next region. The latter technique helps to avoid the internal cancellation of fields that might occur if the entire thorax were magnetized at once.

Another approach to measuring magnetic dust burdens in the thorax has been described by Stern et al. (1986). It uses two matched pairs of Helmholtz coils. The alternating current in the coils is balanced, so the magnetic fields cancel each other midway between them; a flux-gate magnetometer is in this field-free region. Insertion of an object into the region of uniform magnetic field of one pair of coils causes a net magnetic field to be generated in the otherwise field-free region at the detector, and the magnitude of the field imbalance depends on the magnetic moment of the object. Ferromagnetic and paramagnetic materials, such as metals, produce positive signals; diamagnetic materials, such as water, produce negative signals. Measurements involve repeatedly sliding the subject to a position centered between the coils and then out again.

MPG, even in its simplest forms, has demonstrated that populations with occupational exposures to various magnetic dusts have higher average remanent fields than do nonexposed control populations (Table 2-1). But there are some inherent problems. A major one is that the relationship between remanent field and thoracic dust burden is highly variable, even in one person. The accuracy with which measurements of remanent field can be used to measure thoracic burden depends on the validity of the assumptions made about the distribution of magnetic particles in the thorax. Thus, persons of extreme body shape and size or with patterns of deposition and retention different from those assumed could have particle burden estimates that differ substantially from their actual ones. The greatest errors might occur with long-term dust burdens. For example, the field from a point (dipole) source varies inversely as the cube of the distance, and the magnetic-field measurements are much more sensitive to particles in the lung periphery, which is closer to the probe(s), than to those in other areas. Particles translocated toward the pleural surface of the lung in clearance could result in an increase in the observed magnetic field, whereas translocation away from this area could result in a decrease. More studies are needed with MPG and selective erasure by depth to clarify the extent of the artifacts associated with incorrect assumptions about the effects of body size and the distribution of dust in the thorax.

Few people are exposed to a pure magnetic material, or even to a material with an established magnetic moment. Many inhaled dusts contain magnetic components that serve as surrogates or tracers for the material of interest, and the magnetic fraction of the total burden often varies according to the size of the particles. With increasing residence time in the lungs, there can also be selective dissolution or clearance. The former can be a major concern in the analysis of dust retention. When a health hazard from some specific exposure is related to nonmagnetic constituents in the dust, whose presence either is not detected or is masked by magnetic components of the dust, MPG might be of little value for obtaining the retention data of primary interest.

In spite of its limitations, MPG does have important positive aspects. It is totally noninvasive and causes no discomfort or apparent risk to the person being assessed. In addition, there is good agreement between estimates of thoracic burden and both estimates of inhalation and evidence of dust accumulation as seen in x-ray studies (Kalliomaki et al., 1978a,b, 1979, 1981; Cohen et al., 1981; Freedman et al., 1982; Stern et al., 1986). MPG provides a means of actively monitoring the dust burden in exposed people in occupations in which magnetic dusts are used or in cases in which magnetic minerals can serve as tracers of another dust of interest (Table 2-3). It can also provide a means of identifying and monitoring workers who, for whatever reason, might be accumulating dust at an "unacceptable" rate (Freedman et al., 1982).

Respiratory Tract Fluids

Respiratory tract fluids that could

TABLE 2-3 Thoracic Dust Burdens in Humans as Determined by MPG

Population	Burden, mg[a]	References
Arc welders	0-2,000	Kalliomaki et al. (1978b); Freedman et al. (1982); Stern et al. (1986)
Asbestos miners and millers	7.8 (mean)	Cohen et al. (1981)
Coal miners	26-1,440	Freedman et al. (1980)
Iron foundry workers	30-600	Kalliomaki et al. (1979)
Stainless-steel welders	10-1,600	Kalliomaki et al. (1981)
Steel-mill workers	10-200	Koponen et al. (1980)

[a]Expressed as mg Fe_3O_4 equivalents. Variability within one occupation often reflects extent of exposure, e.g., duration of employment.

be monitored for markers of exposure or effects include sputum, nasal-lavage fluid, and bronchoalveolar-lavage (BAL) fluid. Sputum and nasal-lavage fluid can be obtained with relatively noninvasive procedures. BAL enables sampling of alveolar lining fluid, which is in direct contact with or intimately related to cells that are involved in injury or disease. However, the procedure is relatively invasive, requiring that a subject undergo fiberoptic bronchoscopy with some anesthetic. Although the procedure is relatively safe in normal or asymptomatic subjects, substantial morbidity might occur in a person with severe pulmonary disease.

Asbestos and wool fibers have been found in sputum from exposed people. In fact, the ability to detect asbestos bodies in BAL fluid from those exposed occupationally to asbestos is being used diagnostically (De Vuyst et al., 1987). Other inhaled particles, such as coal dust or mineral dust, should be detectable in sputum, nasal-lavage fluid, and BAL fluid from heavily exposed persons (Roggli et al., 1986). The sensitivity of such monitoring and the relationship between what is in the samples and the extent of exposure are unknown.

Recent studies have shown BAL to be a sensitive indicator of effects of environmental pollutants on the lung. Koren and co-workers (1989) reported that, 18 hours after a 2-hour exposure to ozone at 0.4 ppm, the proportion of polymorphonuclear leukocytes in BAL fluid from healthy, nonsmoking men increased by a factor of 8. Similar but smaller increases were seen in immunoreactive neutrophil elastase. Markers of vascular permeability in the BAL fluid doubled. Complement fragment C3a increased by a factor of 1.7, prostaglandin E_2 by a factor of 2, fibronectin by a factor of 6.4, and urokinase plasminogen activator by a factor of 3.6. Smaller exposures (at 0.1 ppm for 7 hours) produced smaller responses, but there were no definite indications of a threshold (above ambient concentration) for ozone-related tissue responses.

The use of such samples as quantitative monitors of exposure will require validation in humans under conditions of known exposure. With the availability of more sensitive means of detecting specific adducts, it might be possible to monitor human samples previously thought to contain nondetectable concentrations of exposure-related material. Pulmonary macrophages represent the cleanup crew of the lung and lower airways and can be expected to reflect the material inhaled and deposited in that area. Sputum, bronchial washes, and BAL fluid contain macrophages. Macrophages are not target sites for tumorigenic responses, but might be used to monitor the extent of recent exposures. Exposures to insoluble particles or fibers should be reflected as phagocytosed material in the macrophages. Particle-associated organic substances are retained in the lung long enough to be metabolized (Sun et al., 1983, 1984). Bond et al. (1984) showed that macrophages can metabolize such compounds as benzo[a]pyrene to reactive substances that could bind to DNA. Newer methods, such as those used by Haugen et al. (1986) to detect benzo[a]pyrene adducts in lymphocytes, could allow detection of DNA adducts in alveolar macrophages. The average time that a macrophage spends in the alveolar space has been estimated at 7-27 days (Van oud Alblas and van Furth, 1979; Bowden, 1983). Thus, alveolar macrophages could be used to measure cumulative exposures over a relative short period (days to weeks). Such an approach deserves further investigation.

Exhaled Air

Exhaled air contains an array of volatile organic constituents that are likely to be in equilibrium with a number of compartments in the lung or can arise from endogenous or absorbed volatile substances circulating in the blood. In addition, some substances in lung air might be in equilibrium with alveolar lining material. Finally, cells within the airspaces (including mucous glands) and cells that are attached to the bronchial epithelium (such as alveolar macrophages) could also contribute to the constituents of lung air. Obviously, exhaled air lends itself to easy noninvasive collection.

Exhaled air can be analyzed with gas chromatography/mass spectroscopy (GC/MS) to yield markers of exposure in the form of volatile substances in the blood. In one case, differences in the contents of exhaled air of a nonsmoking submariner before and after a cruise gave information on the environment of the submarine (Knight et al., 1984, 1985). The halogenated hydrocarbons used for refrigeration on the submarine were easily detectable in the submariner's breath. The exhaled air contained both endogenous metabolites and atmospheric contaminants from the submarine. The comparison between the before-cruise sample and the after-cruise sample helped to distinguish between the metabolites and contaminants. For example, isoprene, the monomeric unit of terpenes known to be an endogenous metabolite in mammals and known to be emitted by a wide range of plants (Tingley et al., 1979), was present in the exhaled breath both before and after the cruise. Conkle et al. (1975) reported the trace contaminants in exhaled air from eight unexposed volunteers. They identified 53 volatile compounds and used a cryogenic trapping system for concentrating trace organic compounds to allow detection of submicrogram quantities. Such a procedure appears to have the sensitivity required for application to environmental exposures.

In the Total Exposure Assessment Methodology (TEAM) study funded by the Environmental Protection Agency (EPA) (Wallace, 1987), the amounts of 11 prevalent volatile organic compounds found in the breath of 355 New Jersey residents were found to correlate with the previous 12-hour average air exposures. That caused the investigators to conclude that "breath measurements may be capable of providing rough estimates of preceding exposures." The same group used analysis of exhaled air to determine exposure to benzene during the filling of a gasoline tank, exposure to tetrachloroethylene in dry-cleaning shops, exposure to chloroform from hot water in the home, and exposure to aromatic compounds in tobacco smoke.

A potential approach that has been little studied is the analysis of exhaled air for volatile metabolites that might be involved in lung disease. The approach is noninvasive and involves sampling of volatile organic compounds from easily obtainable physiologic materials, such as breath and saliva. There are two critical requirements for this type of analysis to be useful: the disease process must lead to the production of volatile metabolites that will be present in exhaled air, and these metabolites must reach measurable concentrations in the total exhaled air. Fulfillment of the latter requirement is limited by the sensitivity and sophistication of the instruments used to analyze the exhaled air. With the use, for example, of gas chromatography combined with mass spectroscopy and appropriate computer analysis, the detectable amount could be as small as several hundred picograms.

Many volatile constituents of body fluids have been characterized in diabetes, respiratory viral infections, and renal insufficiency (Zlatkis et al., 1981). In several diseases, breath analysis with GC/MS has revealed the presence of simple endogenous alcohols, ketones, and amines and numerous compounds of endogenous origin (Chen et al., 1970a,b; Krotoszynski et al., 1977; Simenhoff et al., 1977; Kaji et al., 1978). For example, concentrations of mercaptans and C_2-C_5 aliphatic acids are increased in the breath of patients with cirrhosis of the liver (Chen et al., 1970a,b; Kaji et al., 1978), and dimethyl and trimethyl amines are present in the breath of uremic patients (Simenhoff et al., 1977). A similar approach to the analysis of other volatile metabolites involved in acute and chronic damage to the lung should be explored.

Blood

Inhaled organic compounds enter the blood directly from the lung, and analysis of the blood for an inhaled compound or its metabolites can provide valuable information on recent exposures. Traditional analytic chemistry techniques involving various types of chromatography and spectroscopy have been used to separate and identify compounds. Recently, an innovative vacuum-line distillation technique has been used to separate volatile compounds and their volatile metabolites in blood (Dahl et al., 1984). The approach has proved useful for toxicokinetic studies of such compounds as butadiene (Bond et al., 1986) and isoprene (Dahl et al., 1987)—compounds that have volatile monoepoxide and diepoxide metabolites.

To be able to monitor cumulative exposures over longer periods, one needs a marker that is not cleared from the blood as rapidly as are organic compounds and their metabolites. Newer approaches to biologic monitoring make use of the fact that reactive metabolites of organic compounds can react spontaneously with nucleophilic sites on macromolecules to form covalently bound adducts to DNA, hemoglobin, and other important proteins. Once formed, the adducts are relatively long-lived in the body (compared with the exposure compound or its free metabolites). Some of the materials of greatest interest, mutagens and carcinogens, are classes of compounds that either are reactive electrophiles or can be metabolized to electrophiles. The electrophiles can then bind to nucleophilic macromolecules, such as proteins and DNA. Blood contains large amounts of two proteins, albumin and hemoglobin, with reactive amino and sulfhydryl groups that can interact with electrophiles. Adducts formed with such proteins can be expected to remain in the blood with the same half-time as the proteins. In the case of hemoglobin, the life span of the protein, approximately 4 months, permits detection of cumulative doses from exposures over several months. Albumin has a shorter half-life—approximately 2 weeks. Recent research (Harris et al., 1987; Poirier and Beland, 1987b) has centered around the use of hemoglobin adducts for monitoring.

Hemoglobin adducts formed from reactive metabolites are being investigated for their potential as biologic markers of exposure. Segerback (1983) reported the formation of hemoglobin adducts in mice exposed to ethene or ethylene oxide. The same group has published many studies on the ability of alkylating agents to form hemoglobin adducts and on the value of monitoring persons exposed to such agents by quantification of these adducts (Osterman-Golkar et al., 1976, 1983; Calleman et al., 1978). Tornqvist et al. (1986) used hemoglobin adducts to determine tissue doses of ethylene oxide in cigarette-smokers. They found an increased amount of hydroxyethylation of the N-terminal valine of hemoglobin that correlated with the amount of ethene in cigarette smoke. Pereira and Chang (1981) studied a series of 15 carcinogens and their ability to form hemoglobin adducts in animals. They found that oral exposures of all the carcinogens produced hemoglobin adducts and that the most efficient binding occurred at the lowest doses (0.1 μmol/kg). Green et al. (1984) found the use of hemoglobin adducts promising for monitoring arylamine exposures in humans, on the basis of a study in rats exposed to 4-aminobiphenyl. Shugart and Matsunami (1985) found that hemoglobin adduct formation provided a suitable monitor of exposure to benzo[a]pyrene in mice. Newmann (1984), in a review article, pointed out that hemoglobin adduct formation not only is a more reliable indicator of internal dose, but also indicates a person's capacity to metabolize an inhaled organic compound to a reactive intermediate.

To judge from the results of studies conducted thus far, the use of hemoglobin adducts to monitor exposure to chemicals is promising. The adducts form with a variety of compounds and allow detection of subnanogram amounts of material. The formation of albumin adducts should also be investigated as a potential marker of exposure. Despite the short half-life of albumin, this protein is available in the serum for reaction with reactive metabolites without the need for the metabolites to cross a cellular membrane. In occupational settings, where small exposures can occur daily, an end-of-shift measurement of albumin adducts would yield a sensitive marker of exposure.

Urine

Recent findings have suggested that

biochemical markers in urine can be useful in monitoring the development of nonneoplastic pulmonary diseases. Clearly, such markers would be advantageous if they could be obtained with noninvasive procedures. However, their major drawback is that they reflect events in the lung only indirectly.

Microsomal metabolism of inhaled organic compounds can produce water-soluble metabolites and their conjugates that are excreted in the urine. These compounds are not usually the toxic forms of an inhaled chemical—although exceptions occur (see Chellman et al., 1986)—but their presence in urine can indicate that exposure to a specific chemical has taken place. If pharmacokinetic information is available from animal studies and from physiologic modeling, the total amount of the metabolites excreted in the urine can be used for quantitative estimates of exposure. For example, phenol in urine has long been used to monitor worker exposure to benzene (Teisinger et al., 1955). More recently, DNA adducts have been monitored in urine (Groopman et al., 1985) with monoclonal antibodies to adducts of aflatoxin B_1.

Urine has also been analyzed for mutagenic activity as a monitor of exposure to genotoxic agents (Bloom and Paul, 1981). The relationship between concentrations of urinary mutagens and risk of cancer is still unknown and requires further research. Analysis of urinary mutagens shows promise, on the basis of such work as that of Camus et al. (1984) in which two strains of mice with different susceptibilities to development of cancer were treated with benzo[a]pyrene and the urinary mutagen concentrations were correlated with tumor formation. In general, the ability to produce urinary mutagens corresponded to susceptibility to tumorigenesis.

The use of urinary mutagens to monitor environmental exposure will be difficult, because such confounding factors as diet, smoking, and occupational exposure introduce uncertainties in interpretation of the assays. Ohyama et al. (1987) reported that ingestion of cooked salmon increased urinary mutagens, whereas ingestion of cooked vegetables did not. Kawano et al. (1987) reported a correlation between increases in mutagens in smokers' urine and the number of cigarettes smoked per day. Similar results were reported by Mohtashamipur et al. (1985). T. H. Conner et al. (1985) reported no increase in urinary mutagens in autopsy-service workers exposed to formaldehyde, but there was an increase in a heavy smoker in the control group. Steel workers exposed to coke-oven emission (benzo[a]pyrene concentrations, 0.01-0.6 $\mu g/m^3$) had slightly higher concentrations of urinary mutagens than unexposed controls, but smoking habits were the major influence on the concentrations (De Méo et al., 1987).

Turnover of extracellular matrix of the lung has been suggested as accompanying tissue remodeling after exposure to environmental pollutants. Some investigators believe that measurement of hydroxyproline or hydroxylysine in urine could reflect collagen turnover; both these amino acids are found only rarely in molecules other than collagen. Kelly et al. (1986) tested the utility of connective tissue breakdown products as markers of current injury after brief exposure of Fischer rats to NO_2. They found a linear increase in hydroxylysine excretion with increasing NO_2 concentration.

Several environmental contaminants can now be monitored in urine. For example, Enterline et al. (1987) have examined exposure to arsenic in men working at a copper smelter in Tacoma, Washington. They found a good correlation between urinary concentrations of arsenic in workers (as a marker of exposure) and respiratory tract cancer.

In clinical studies by Hatton et al. (1977), hydroxyproline glycosides were increased in the urine of three American astronauts who were accidentally exposed to toxic fumes of NO_2 during the descent phase of their mission. More recently, Yanagisawa et al. (1986) attempted to correlate NO_2 exposure with urinary hydroxyproline-to-creatinine ratios in 800 women who were mothers of primary-school children in two communities around Tokyo. The ratio was found to be correlated with NO_2 exposure and with numbers of cigarettes smoked actively and passively. Such experiments appear promising, but methods must be developed to allow identification of specific sources (e.g., lung, liver, and bone) of the breakdown products, inasmuch as collagen is present in many organs other than the lung. Studies to define site specificity of the breakdown products are essential.

Adipose Tissue

Organic vapors and gases that are inhaled can be retained in the fat depots of the body long after they have cleared from other parts of the body. Biopsies of fat could potentially be used to obtain information on lipid-soluble compounds to which a person has been exposed.

An attempt to use that approach to determine which Vietnam veterans had been exposed to Agent Orange was based on laboratory animal studies that showed that the toxic contaminant of Agent Orange, 2,3,7,8-tetrachlorodibenzo-p-dioxin (TCDD), accumulates in fat. Four groups of male volunteers were included: five "heavily exposed" veterans, 20 men who believed that they had been exposed to Agent Orange in Vietnam, 11 who had not been in Vietnam and had had no other contact with Agent Orange, and three Air Force officers who had definitely worked with either Agent Orange or TCDD (Gross et al., 1984). Samples of 10-30 g of adipose tissue were taken from the abdominal wall of each volunteer and analyzed with gas chromatography and high-resolution mass spectrometry. The results indicated that it was possible to detect TCDD in human fat, but it was present in the fat of both the control (supposedly nonexposed) men and the veterans. Four of the five heavily exposed veterans had a mean concentration of TCDD in fat of 55 ± 34 parts per trillion (ppt), and one did not have detectable TCDD. The mean in the other Vietnam veterans was 4 ± 1 ppt, and the mean in the control subjects was 6 ± 3 ppt.

The 2,3,7,8-tetrachlorodibenzo-p-dioxin concentrations in the adipose tissue of Missouri residents distinguished between persons with a history of exposure to the chemical and control (presumable nonexposed) persons (Patterson et al., 1986). Anderson (1985) stressed the importance of analyzing blood samples and adipose tissue samples at the same time to provide more information on the partitioning between the two compartments. When sufficient information of this type is available, blood samples can be used to predict concentrations of the compound in fat.

Some success has been achieved in monitoring adipose tissues for evidence of exposure to polychlorinated biphenyl (PCB) congeners and for exposure to dioxins and furans in accident situations, in which concentrations are higher than would be expected in ordinary environmental exposures. M. S. Wolff et al. (1982) reported concentrations of PCBs in plasma and adipose tissue that were related to duration and magnitude of exposure in persons occupationally exposed to PCBs. Schecter et al. (1985) examined persons 1-2 years after exposure to PCBs, dioxins, and furans in the Binghamton, N.Y., State Office Building incident and found PCBs in their blood.

The National Adipose Tissue Survey of the Environmental Protection Agency's National Human Monitoring Program (Lucas et al., 1982) is an example of the excellent use that can be made of tissue banks for retrospective studies of the influence of occupation, geographic location, age, and sex on the concentrations of halogenated hydrocarbons in human fat.

DNA and Protein Adducts

The use of DNA and protein adducts as a measure of exposure and of risk of tumor formation is under intense investigation (Poirier and Beland, 1987a). A rapid, sensitive method for detecting adducts has aided research (Randerath et al., 1985), but does not distinguish between types of adducts. The development of monoclonal antibodies to specific DNA adducts promises to provide valuable monitoring tools for the future. Monoclonal antibodies have been developed for such specific DNA adducts as those formed from benzo[a]pyrene diol epoxide, 1-aminopyrine,

8-methoxypsoralen (Santella et al., 1987), and the carcinogen aflatoxin B_1 (Groopman et al., 1987).

Some of the most extensive work has been done on adducts formed after exposure to aflatoxin B_1 (Groopman et al., 1987). Studies have shown that the concentration of DNA adducts formed is quantitatively related to the intake of the carcinogen. In addition, the kinetics of the removal of the adducts have been determined, and a monoclonal antibody to the adducts has been made and can be used to measure them in DNA and in urine. A constant proportion of DNA adducts appears in urine as a function of time after exposure (Groopman et al., 1985). Such work to determine quantitative relationships between exposure and adduct formation is needed for other carcinogens in the environment.

Recent advances in analytic techniques allow the detection of DNA adducts in lymphocytes. Haugen et al. (1986) reported the determination of polycyclic aromatic hydrocarbons (PAHs) in urine, benzo[a]pyrene diolepoxide-DNA (BPDE-DNA) adducts in lymphocyte DNA, and antibodies to the adducts in serum of coke-oven workers. To measure the adducts in lymphocytes, an ultrasensitive enzymatic radioimmunoassay and synchronous fluorescence spectrophotometry were used. Approximately one-third of the workers had detectable BPDE-DNA adducts in their lymphocytes and antibodies to epitopes on BPDE-DNA adducts in their serum. The concentrations of PAHs in the atmosphere of the workplace were 212-315 $\mu g/m^3$.

The problems and promise of the use of macromolecular adducts (both protein and DNA adducts) in biologic monitoring were reviewed at a recent workshop (Poirier and Beland, 1987b). The long-term hope is to be able to relate the concentration of adducts to the extent of exposure and to the risk of tumor formation. To achieve that, animal experimentation must indicate the relationship of adducts to the extent of exposure and tumorigenesis in various exposure regimens, and then the relationship for humans must be validated by adduct analysis of human tissues under known exposure conditions. In other words, the pharmacokinetics of adduct formation must be known. Work addressing the rate of accumulation and persistence of DNA adducts, such as that reported by Belinsky and Anderson (1987) on the accumulation of 4-(N-methyl-N-nitrosamino)-1-(3-pyridyl)-1-butanone (NNK), is required. A linear relationship between exposure dose and adduct formation (with both protein and DNA) has been shown in single-exposure experiments for hemoglobin adducts to 4-aminobiphenyl (Green et al., 1984), to alkylating agents (Osterman-Golkar et al., 1976, 1983), and to 4-dimethylaminostilbene (Newmann, 1984) and for DNA adducts to benzo[a]pyrene and to aflatoxin B_1 (Pereira et al., 1979; Appleton et al., 1982; Dunn, 1983; Adriaenssens et al., 1983). With continuous exposures to alkylating agents (Swenberg et al., 1986), adduct concentrations increase with time and eventually reach a plateau or equilibrium where the rate of formation equals the rate of removal.

Poirier et al. (1987) reported information that can aid in validating adduct formation in humans. DNA adducts with the cancer chemotherapeutic agent cisplatin were measured in the nucleated peripheral blood cells and other tissues of patients receiving the drug. A positive correlation was observed between adduct concentrations in the blood cells and cumulative dose of the drug over several months. A major need that remains is to identify the adducts that have biologic significance. Further research in animal models or in human clinical work, such as that reported by Poirier et al. (1987) in cancer patients on chemotherapeutic regimens, is required to correlate adduct formation with biologic effects.

DNA adducts have been detected in respiratory tract tissues, and the amount of these adducts increases after inhalation exposure to some organic compounds. DNA adducts identified as BPDE deoxyguanosine adducts were detected in the lungs of rats exposed to BaP (Wolff et al., 1989). In another study, the concentration of total DNA adducts in various regions along the respiratory tract was compared with the site of tumor development in rats exposed to diesel exhaust. The concentration of

DNA adducts, detected with the ^{32}P-post-labeling technique, was highest in the peripheral lung tissue, the site of tumors in rats chronically exposed to the exhaust (Bond et al., 1988). Such results point to the importance of DNA adducts as measures of effective dose of inhaled carcinogens. However, it will be important to understand the kinetics of formation and repair of individual adducts (as opposed to total adducts), to elucidate the relationship of DNA-adduct formation to carcinogenesis.

Recent work has shown that adducts formed with protamine, a basic protein associated with sperm DNA, might also be potential markers of exposure at the critical site for adverse biologic effects. In one of the first reports linking adduct formation with a specific deleterious effect, Sega et al. (in press) identified one of the protamine adducts formed after acrylamide exposure as S-carboxyethyl-cysteine. The formation of that adduct acts to break S-S bonds in the protamine and might be the basis for the chromosomal breakage induced by acrylamide.

MATHEMATICAL MODELING OF EXPOSURE

Detailed toxicokinetic studies in humans are usually impossible, because of ethical constraints and because not all relevant human organs and tissues can be sampled. However, mathematical modeling of the disposition and fate of inhaled chemicals in animals is useful for extrapolating data between species (NRC, 1987). A strong impetus for the development of mathematical models of the deposition and retention of inhaled particles came with the nuclear age. The Task Group on Lung Dynamics of the International Commission on Radiological Protection developed such a model to determine the dosimetry associated with inhaled radioactive particles (Task Group on Lung Dynamics, 1966). That model, which remains the basis for modeling the toxicokinetics of inhaled insoluble particles today, predicts the dose to internal tissues throughout the respiratory tract and allows extrapolation of results from animals to humans.

Mathematical modeling has also been used successfully to predict the uptake, distribution, and elimination of inhaled lipid-soluble volatile materials (Andersen, 1981; Fiserova-Bergerova, 1983; Andersen and Ramsey, 1983). Such models are useful for calculating doses to critical tissues, for extrapolating between species (including humans), for assessing hazards (F. J. Miller et al., 1987), and for guiding research. Recent advances in modeling that include physiologic characteristics—such as blood flow into and out of an organ, membrane permeability, and chemical partitioning among blood, other tissues, and air—allow rather accurate extrapolations between species (Fiserova-Bergerova and Holaday, 1979; Fiserova-Bergerova et al., 1980; Andersen, 1981; Fiserova-Bergerova, 1983). A pharmacokinetic model developed and validated in animals can be adjusted for the physiologic characteristics appropriate for humans and validated by analyzing readily available human material, such as blood or excreta from humans exposed in clinical studies or in other situations of known exposure (Reitz, 1986).

An example of that approach is the work of Ramsey and Andersen (1984), who developed a physiologically based pharmacokinetic model of the behavior of inhaled styrene in rats to predict accurately the behavior of inhaled styrene in humans. Using a set of physiologic and biochemical constants, the investigators were able to simulate the behavior of inhaled styrene in rats. Experimentally determined values for several measures were used: body weight, alveolar ventilation, blood flow rates, tissue volumes, blood-air partition coefficient, tissue-blood partition coefficients, maximal reaction rate (V_{max}), and the Michaelis constant (substrate concentration at half the maximal reaction rate). They then extrapolated the values to humans and found that the model accurately predicted the amount of styrene that had previously been published to be in blood and exhaled air of humans exposed in clinical studies (Ramsey et al., 1980). The same group (Andersen et al., 1987) did a similar study later with methylene chloride and for the first time extended the model to include values to take into account the metabolic capacity of the lung, as well as the liver. This type of modeling constitutes a powerful tool for extrapolating from animal to human data.

Some precautions should be mentioned. In the animal studies, it is important to determine how and at what rate a chemical and its metabolites are cleared from the body in different doses or exposure regimens, to find the range of doses over which the disposition and metabolism of the chemical are linearly related to dose. Such information is required for extrapolation from animal studies (normally at high doses) to the low doses normally encountered by humans. It is also important to determine the effect of repeated exposure on the fate of a chemical. Repeated exposure at low concentrations, the most commonly encountered human exposure regimen, could induce enzymatic changes that affect the toxicity of a chemical. Such studies can be conducted in animals.

The biggest problem in using animal toxicokinetic data for human risk assessment is the potential for species differences in metabolism. Recent work by Medinsky et al. (in press a,b) extended physiologic modeling to include the disposition of both the parent substance and its metabolites in rats and mice given benzene orally or by inhalation. To extend such models to humans, however, requires some information on human metabolism of the chemical of interest. For example, rats and mice metabolize benzene differently (Sabourin et al., 1988). Extension of the model of Medinsky et al. to humans by changing the physiologic parameters from those of the rat and mouse to those of humans necessitates choosing the appropriate metabolic parameters for humans. If one uses either the rat or mouse metabolic parameters, the extension of the model will predict only how a very large rat or a very large mouse would handle benzene. Therefore, it is essential to have enough information on the human metabolism of a chemical to permit a valid extension of a physiologic model from animals to humans. Comparison of metabolism of xenobiotics by liver slices or cultured cells from laboratory animals and humans should aid in making such extrapolations.

CLINICAL TECHNIQUES FOR GATHERING DATA

Neither the clinical history of a patient nor information obtained on a population with a standardized questionnaire constitutes biologic material in an ordinary sense. However, both can play a role in identifying markers of exposure. Obviously, age is an important biologic determinant of disease risk, and the simplest and most efficient way to determine it is to ask. Similarly, symptoms related to altered biologic states that indicate likelihood of disease can be simply asked about. For example, severe respiratory illness before the age of 2 years implies risk of lower respiratory illness at 6-11 (Samet et al., 1983), and persistent wheezing during childhood predicts diminished pulmonary function in later life (Weiss et al., 1980).

Questionnaire

Since the early 1950s, efforts to develop standardized procedures for gathering clinical and epidemiologic data on pulmonary health status have been in place (Samet, 1978). The efforts have been directed toward reducing bias and ensuring reliability and validity of the information obtained. The first recognized standard questionnaire became available in 1960: the British Medical Research Council (BMRC) questionnaire on respiratory symptoms. The questionnaire was moderately revised in 1966 and 1976. The American Thoracic Society (ATS) adapted the BMRC questionnaire in 1968 and published it with instructions for its use in the United States. Additional modifications have taken place, and the questionnaire has been translated and used extensively throughout the world.

In 1978, after an extensive evaluation of the questionnaire, ATS and the Division of Lung Disease (DLD) of the National Heart, Lung and Blood Institute recommended a new version (Ferris, 1978). An extensive review of the 1978 ATS-DLD questionnaire is beyond the scope of this report, but we should note that it is effective and reliable for ascertaining respiratory symptoms related to the ill effects of cigarette-smoking or other respiratory irritants. Because smoking-habit information is obtained in a standardized format, the questionnaire makes it possible to measure current and lifetime exposure and allows for comparisons between groups.

Specific biologic risk factors identified in the questionnaire for chronic respiratory diseases are summarized in Table 2-4. Other questions can also be used in a clinical setting with patients and provide useful measures of the severity of disease, but are not as well standardized as those discussed. With the use of standardized questions, population groups with different degrees of risk can be defined; the questions can then become useful to delineate biologic markers.

Although generally considered effective, the use of questionnaires clearly has limitations. Repeated assessments in what are thought to be stable populations are not without variance, and few (if any) questions have been independently validated (Samet, 1978). The original survey questions in the ATS-DLD questionnaire were designed specifically to identify smoking effects. Few questions are related to specific environmental agents other than cigarette smoke. Now that smoking occurs only in a minority of

subjects, alternative questions might be warranted. And concern has been expressed that better questions need to be developed to deal with symptoms of reactive-airway disease, such as asthma.

The present questionnaire includes only one question on asthma and one series of questions on wheezing. Developing useful questions on reactive-airway disease will require correlations of responses to newly constructed questions with a readily acceptable physiologic test of increased airway reactivity (e.g., nonspecific airway hyperreactivity).

Clinical signs and examinations by physicians or other trained observers usually have not proved particularly useful as biologic markers of predisease status. The yellow-stained fingers of a chronic cigarette-smoker might be just as useful "biologic markers" as are questions about smoking habits.

Lung Sounds

One item in a physical examination that is potentially useful as a biologic marker is the recording of lung sounds (Wooten et al., 1978). However, instruments for reproducible scaling of lung sounds remain to be developed, and credible scientific investigation of the usefulness of lung sounds has been sparse (Loudon and Murphy, 1984). Existing technology allows accurate recording of sounds transmitted from the airways and parenchyma to the chest wall. Integrating and possibly automating

---

TABLE 2-4 Biologic Questionnaire Data That Can be Used to Predict Chronic Respiratory Disease in Adults

Risk Factor	Comments
Age	Few cases of emphysema below age 40; risk increases with age; greater risk in men (might be related to greater exposure)
Chronic cough and/or phlegm (chronic mucus hypersecretion)	Associated with cigarette-smoking; not independently associated with risk of COPD; might be associated with excess risk of lung cancer
Episodes of cough and phlegm	Associated with chronic mucus hypersecretion, increased episodes of pneumonia, and time lost from work
Persistent wheeze	Increased risk of asthma; reduced pulmonary function; increased susceptibility to pulmonary irritants
Dyspnea	Poorly correlated with pulmonary function, but at severe grades associated with reduction and inability to perform pulmonary function test adequately
Smoking history	Important predictor of risk (e.g., of lung functional impairment and lung cancer)

the information obtained from those sounds would allow useful categorization of early respiratory injury that could be correlated with other markers of exposure. The recording of lung sounds has the advantage of being noninvasive and warrants further exploration.

An example of the use of lung sounds is shown in a survey of 270 asbestos factory workers. Shirai et al. (1981) found fine discontinuous lung sounds (i.e., crackles) more frequently in asbestos workers (32.2%) than in controls (4.5%). There was good agreement between findings on chest auscultation and sound recordings. In fact, it was found that bilateral basal crackles occurred in asbestos workers before radiographic abnormalities were present. Fine crackles might be valuable as an early diagnostic marker of pulmonary asbestosis.

Respiratory Function

The use of respiratory function studies is described in detail in the next chapter. Nonetheless, the summary report by Lippmann (1988) related to acute ozone exposures provides an example of the use of clinical techniques to assess exposure in epidemiologic studies. He concluded that there were "progressively increasing functional decrements with each consecutive hour of O_3 at 0.12 ppm, as well as substantial increase in bronchial reactivity. . . ." Information presented at the U.S.-Dutch meeting "suggests that lung inflammation from inhaled O_3" has no threshold down to ambient background concentrations. It was further concluded that rats constitute a good test model for the observed human response to ozone, even though they are less sensitive than humans.

Studies in healthy and asthmatic adolescents (Koenig et al., 1987) used standard measures of respiratory function—e.g., peak flow, respiratory resistance, and forced expiratory volume (FEV)—and found significant increases in respiratory resistance in both healthy and asthmatic adolescents after exercise exposure to ozone at 0.18 ppm, but no differences in response between the two groups. Koenig reported that the most important finding

is that there was little difference in the effects of ozone or nitrogen dioxide between healthy and asthmatic subjects.

Other studies of respiratory responses to ozone exposure in healthy, active children have also used standard respiratory function measures and found highly significant changes in PEFR in response to changes in ambient ozone concentrations (Spektor et al., 1988).

Imaging

Other noninvasive techniques that could be regarded as yielding biologic markers of exposure or disease include radiography and other imaging techniques. Unfortunately, the techniques seldom provide evidence of specific exposure or disease. Two notable exceptions are the pneumoconioses that constitute specific evidence of dust accumulation in the parenchyma and the pleural reactions that are produced by asbestos exposure; in both cases, the findings must be accompanied by appropriate exposure data if they are to be reliable.

The major rationale for obtaining screening x-ray pictures has been to identify persons with clinically silent pulmonary tuberculosis. More recently, chest x-ray examination has been used to screen asymptomatic smokers for lung tumors and to screen for occult lung or heart diseases in general hospital admissions. Clinicians have recommended the use of chest x-ray pictures to establish a baseline for comparison, especially in people presumed to be at risk of lung disease. Even in such select groups, the merits of screening have been debated. Although enthusiasm for chest x-ray pictures for screening, in general, seems to be waning, it is worth emphasizing that radiologic examination remains a major diagnostic tool for revealing occupationally induced interstitial lung disease.

In the 1970s, the International Labor Organization (ILO) produced a standardized procedure for obtaining and reading chest x-ray pictures that allowed crude measurement of exposure to mineral dust (Jacobson and Lainhart, 1972). Refinements of the procedure in the 1970s have led to a standardized method that increases the uniformity and reproducibility of readings of these films made by certified persons (ILO, 1980). At least in serious cases, the degree of parenchymal infiltrate is correlated with histopathologic evidence of dust accumulation (Seaton, 1983). Thus, the x-ray pictures in specific circumstances become the documentation of the biologic marker of inorganic dust exposure. For example, although it is less quantitative, the appearance of pleural plaques or pleural thickening on a chest x-ray picture in the presence of a history of asbestos exposure implies a greater likelihood of future asbestos-related disease.

Sophisticated imaging techniques used in clinical practice, such as computed axial tomographic (CAT) scanning, can locate and characterize small lesions that could be considered biologic markers of disease. However, the procedures are clinical diagnostic tools and not likely to be used in screening populations of apparently healthy subjects.

Such techniques are of tremendous importance to experimental medicine and toxicology, because they permit resolution at the subcellular level and might ultimately promote the identification of biologically effective dose, early biologic effect, and altered structure or function.

SUMMARY

The development of biologic markers of exposure to xenobiotics offers much promise. New molecular biologic techniques permit the measurement of such molecular markers as adducts formed with macromolecules in the body; the techniques can be used to detect adducted material in blood, urine, and tissue samples and are sensitive enough for the measurement of adducts formed with DNA or protein in cells washed from the respiratory tract or collected in sputum.

Innovative procedures, such as magneto-pneumography, allow estimation of the lung burdens of some types of particles in the lung. Refined histologic techniques have revealed the cellular sites of deposition of inhaled particles in the lung and thus created the potential for calculating the dose to critical cells. Techniques for analyzing markers are well advanced; e.g., new techniques allow analysis of exhaled air, sputum, nasal lavage fluid, and bronchoalveolar lavage fluid for chemical evidence of exposure to specific pollutants.

Mathematical modeling has advanced to the point where models now include physiologic measurements, such as blood flow rates, ventilation rates, metabolic rates, and both blood-air and blood-tissue partition coefficients. The models have made it much easier to extrapolate from animal pharmacokinetic data to predicted disposition in humans.

To make optimal use of the new techniques, we must determine the relationship between markers of exposure and the characteristics of the exposures that generate them. The markers usually yield only yes-no answers; that is, a particular exposure did or did not occur. But we need to determine the kinetic relationships between formation and breakdown of markers, so that we can use mathematical models to answer the question, "Given this amount of this marker in this tissue, what exposures could have produced the marker?" In addition, there is a need to explore such readily available respiratory tract fluids as nasal fluids and sputum for new chemical markers of exposure to specific pollutants.

Finally, we must determine the mechanisms by which environmental pollutants induce lung disease. What are the sites of toxic actions? How much of a given pollutant is required at a given site to produce a given toxic response? Knowledge of the mechanisms by which toxicity occurs should provide the most pertinent information on potential early markers of exposure to environmental pollutants and initial stages of response to them.

3

Markers of Physiologic Effects in Intact Organisms

This section discusses markers of pulmonary response that can be applied in studies of intact human subjects (all can also be applied to animals). The focus is on physiologic tests of respiratory system function that reflect early biologic effects or underlying changes in lung structure, function, or sensitivity to inhaled materials.

Most of the techniques reviewed here can be applied in a relatively noninvasive manner and do not require anesthesia, catheterization, or collection of tissue samples. Many are suitable for epidemiologic studies of large populations and have the characteristics of mobility of equipment, short subject interaction time, minor requirement for subject training, ability to be performed by technicians, and automated data processing; an example of such a technique is spirometry, the most common of those discussed in this section. The equipment, personnel, and sample-collection requirements of others imply that they are likely to be used only in stationary facilities. Many centers are suitable for these tests, but they are most appropriate for evaluation of small populations. Techniques that use gamma-camera imaging are examples of this class because gamma-camera equipment is large. Other approaches require equipment and expertise that are likely to be present only in a few specialized laboratories, for example, measurements of the dispersion or clearance of inhaled boluses of particles.

The markers of early biologic effects or altered structure/function discussed here are in four general categories. The first includes markers of respiratory or gas-exchange function of the lung. These are derived from tests of ventilation and its control, lung mechanical properties, intrapulmonary gas distribution, and alveolar-capillary gas exchange. A wide variety of such tests have been developed and are in common use to evaluate lung functional competence. Although all these tests yield markers of lung response, only a few thought most likely to have potential for detecting responses of populations to environmental exposures are discussed in detail. The second category includes markers of increased airway reactivity, both to specific environmental agents and to standardized physical or pharmacologic bronchial provocation. Although these are usually indexes of respiratory function, they are distinct from the first category, in that the focus is not on gas exchange. The third category includes markers derived from measurement of the clearance of particles. These measurements examine an important set of defense mechanisms of the respiratory system, and alterations of the mechanisms sometimes give an early indication of an adverse impact of inhaled environmental agents. The fourth category includes markers of increased permeability of the air-blood barrier, which is sometimes an early feature of lung injury due to inhaled materials.

The usefulness of assays of physiologic function is generally limited to their ability to demonstrate responses to inhaled environmental agents. They reveal the functional manifestations of structural changes in the respiratory system, whether the changes are transient (e.g., bronchoconstriction) or lasting (e.g., fibrosis). As a group, the assays have little specificity for specific environmental agents. The lung responds to injury in a limited number of ways, and the functional manifestation of a given type of injury is often the same, regardless of the causative agent. An exception to both those generalizations might be the usefulness of an increase in airway sensitivity to specific agents as a marker of exposure and sensitization.

The sensitivity of physiologic tests varies, but many demonstrate substantial intersubject variability. The sensitivity is often improved when control populations are studied or when individual baseline values are obtained; however, there is often a reliance on predicted normal values. The results of physiologic assays generally depend on technique, and both the degree of standardization of technique and the magnitude of the data base for predicting normal values vary widely from assay to assay. Some of the physiologic tests offer the advantage that a response, once detected, can be placed in a context that can be interpreted in terms of its practical impact on the subject. That is based on knowledge of correlations among measured values related to function, subjective perception of ill health, clinical lung disease, and impairment of physical performance.

RESPIRATORY FUNCTION

The respiratory functions of the lung include mechanisms involved in ventilation, gas distribution, alveolar-capillary gas exchange, and perfusion. Although developments in the understanding and evaluation of respiratory function are continuing, most of these physiologic phenomena are well-studied, and many assays of function are long established. Table 3-1 lists several common assays of respiratory function. Note that each listed category contains numerous individual tests or measured characteristics.

Only a few of the many assays of respiratory function are discussed in detail in this section, but all the assays listed are useful in a clinical setting and yield potential markers of response to environmental exposures. As there is considerable literature on the performance and interpretation of the assays; it would be inappropriate to review it all here. The intent of this review is to comment on the function tests that the Committee thought were most likely to have potential for studies of the effects of environmental exposures in occupational groups or more general populations.

Considerable attention has been focused on the development of respiratory function tests that are sensitive to alterations in the region of terminal bronchioles and respiratory bronchioles—i.e., small-airway disease. That anatomic region is at particular risk from several types of inhaled toxicants. Therefore, respiratory function tests that have been proved (or proposed) to have particular sensitivity to small-airway disease are discussed here. Current knowledge suggests that a given test has little usefulness for producing markers of effects of specific environmental exposures.

Spirometry

The forced expiratory maneuver, which records the time taken to expel as quickly as possible as much gas as possible from a full deep inspiration, is the mainstay of both clinical pulmonary physiology and epidemiologic field studies designed to

TABLE 3-1 Assays of Respiratory Function

Breathing pattern
    Respiratory frequency, tidal volume, and minute volume
    Inspiratory-expiratory times and flow rates
    Alveolar ventilation
Physiologic subdivisions of lung volume
    Total lung capacity
    Vital capacity
    Functional residual capacity
    Residual volume
    Inspiratory and expiratory reserve volumes
Spirometry (forced exhalation)[a]
    Forced vital capacity (FVC)
    Forced expired volume in 1 second (FEV_1) and %FVC in 1 second
    Peak expiratory flow
    Mean midexpiratory flow
    Flows at selected absolute lung volumes or portions of FVC
Breathing mechanics
    Dynamic lung mechanics
        Dynamic lung compliance[a]
        Total pulmonary resistance
        Airway resistance and conductance, and specific airway
            resistance and conductance
        Oscillation mechanics (respiratory system impedance,
            composed of compliance, resistance, and inertance)[a]
    Static-quasistatic lung compliance[a]
Intrapulmonary gas distribution
    Single-breath gas washout[a]
    Multiple-breath gas washout
    Imaging of radiolabeled gas and particles[a]
    Particle bolus distribution[a]
Alveolar-capillary gas transfer
    Blood gases, pH, and alveolar-arterial gas tension differences
    Oxygen and carbon dioxide exchange at rest
    Diffusing capacity for carbon monoxide[a]
    Gas exchange during exercise[a]
    Multiple-gas evaluation of ventilation-perfusion relationships
Evaluation of respiratory control

[a]Discussed in this report.

explore the development of chronic obstructive pulmonary disease. The volume of air expelled in the first second is termed the forced expiratory volume in 1 second (FEV_1). The total amount expelled is called the forced vital capacity (FVC). The flow-volume curve is a plot of expiratory flow rate against expired volume, and it is also analyzed to evaluate flow limitation.

Patients with chronic obstructive respiratory diseases are those whose airflow limitation prevents them from continuing activities that they would otherwise be able to perform (Speizer and Tager, 1979). There is a continuum between the normal state and the diseased state. Most people would consider themselves substantially disabled when their FEV_1 approached 40-50% of the predicted value. Most physicians would take an FEV_1 below 65% of the predicted value as indicative of obstructive disease.

FEV_1 tends to decline smoothly in adult life with modest acceleration with increased age (Figure 3-1) (Speizer and Tager, 1979). Several studies have shown that airflow function is a strong predictor of morbidity (Fletcher et al., 1976). Thus, FEV_1 is a good biologic marker of risk of developing obstructive pulmonary disease.

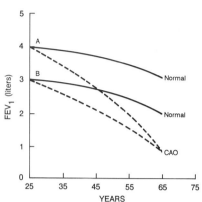

FIGURE 3-1 Decline of FEV_1 at normal rate (solid line) and at accelerated rate (dashed line). "A" represents a person who has attained a "normal" maximal FEV_1 during lung growth and development. "B" represents a person whose maximal FEV_1 has been reduced by childhood respiratory infection (CAO). Reprinted with permission from Samet et al., 1983.

Additional measures of ventilatory function that are general biologic markers of risk of chronic respiratory disease are based on a forced expiration. To use them, one must have recording devices that capture the essential components of the maneuver or the entire curve. The total volume of an expiration, the FVC, has been recognized since the mid-nineteenth century as a predictor of "vital status" (Hutchinson, 1846). In the 1960s, the Framingham heart program found FVC to be a predictor of total mortality (Kannel et al., 1980). More recently, FEV_1 and FVC measurements obtained some 20 years previously were used to predict specific respiratory disease mortality (Peto et al., 1983). Other components that are believed to represent the flow characteristics of the airways include flow after 50 or 75% of the volume has been forcibly expelled (forced expiratory flow of 50 or 75%, FEF_{50} or FEF_{75}) and the maximal midexpiratory flow (MMEF, the slope of the line drawn between points at 25 and 75% of the FVC).

The same measures have all been used (with other measures discussed below) to test the effects of exposure to an array of environmental agents, generally at concentrations that exceed only slightly those occurring in the environment. In each case, pre-exposure, postexposure, and recovery measurements were compared. The interpretation of those measures as biologic markers of risk is only partly

understood. When asthmatic subjects are exposed at rest to SO_2 at 0.25-0.5 ppm, some undergo significant reductions in FEV_1 (Sheppard et al., 1980). Whether the results predict which asthmatics are more susceptible to naturally occurring environmental insults is as yet unknown. However, we do know that children who report wheezing or asthma generally have more respiratory symptoms than those who do not when exposed to ambient environments with particulate pollution (Ware et al., 1984). No relation between pollutant concentration and magnitude of FEV, FVC, or MMEF change has been discerned, but children with a history of wheezing clearly have lower MMEF.

In most chronic respiratory diseases other than asthma, lost ventilatory function does not return. By the time impairment is judged to be significant, there is little need for subtle biologic markers of risk of disease (it might still be appropriate to use markers to study mechanisms). Therefore, it is important to consider how the sensitivities of the other markers compare with the sensitivity of FEV_1. For example, subtle decreases in flow at 50 or 75% of vital capacity (FEF_{50} or FEF_{75}) might be associated with cigarette-smoking without necessarily being simultaneously associated with reduced FEV_1. Some have argued that those findings represent changes in small airways (Gelb and Zamel, 1973). Typically, reduced FEF_{50}

or FEF_{75} without a reduced FEV_1 has not been used to define obstructive airway disease (Speizer and Tager, 1979). MMEF similarly might be reduced in association with exposure to respiratory irritants, but this has not typically been used to define obstructive disease without a reduced FEV_1. Those more subtle measures are useful as biologic markers, because they might indicate earlier or more subtle damage to small airways that, if not reversed, could lead to more severe and irreversible damage reflected in reduction in FEV_1 and eventually in FVC.

Mechanical Properties of the Lung

Dynamic Lung Mechanics

Measurement of dynamic lung mechanics is a means of assessing the work of breathing. The work of breathing is incurred in the need to overcome elastic, resistive, and inertial forces of the lung tissue and air column (Mead and Agostoni, 1964; Rodarte and Rehder, 1986). Dynamic lung mechanics—lung mechanical properties during breathing—are usually expressed in terms of dynamic lung compliance (indicative of work required to stretch the lung) and airway resistance (indicative of work required to overcome resistance to airflow). Classical measures of dynamic lung mechanics are often useful in evaluating clinical lung disease, but by themselves have low potential as sensitive markers of lung response to environmental exposures.

Tests commonly applied in epidemiologic studies to detect abnormalities of respiratory function typically evaluate lung volume and resistance to airflow, but do not examine the compliance. Lung compliance is reduced in disorders such as inflammation and fibrosis, which increase lung elastic recoil, and is increased in diseases like emphysema, which decrease elastic recoil. Measurement of static or quasistatic compliance demonstrates those changes more sensitively than does measurement of dynamic compliance. Aside from forced-oscillation methods, measurement of dynamic compliance requires placement of an esophageal balloon catheter.

Although that is not difficult or hazardous, it is sufficiently time-consuming and unpleasant for its use to be limited usually to selected clinical subjects and to the physiology laboratory. Oscillation techniques now constitute the most likely use of lung compliance as a marker of response.

Dynamic lung compliance depends on breathing frequency. Compliance decreases with increasing breathing frequency, because of regional inhomogeneities among lung units (Otis et al., 1956). Measurement of the frequency dependence of compliance was introduced as one of the first tests "specific" for small-airway disease (Woolcock et al., 1969). Compliance is measured as the subject breathes over a range of frequencies, and the magnitude of the reduction is noted. Although the dependence of compliance (or resistance) on frequency is often mentioned as a potential marker of small-airway disease, these changes have actually been correlated with structural changes in the lung in only a few studies (Berend, 1982). The dependence on frequency has been shown to be abnormal in a large portion of asymptomatic young smokers (Martin et al., 1975). The compliance test is not broadly used, and there are few data on which to base either an estimate of its usefulness in population studies or estimates of normal values.

The most common method of assessing resistance to airflow is spirometry during forced exhalation. Resistance to airflow during either forced exhalation or tidal breathing is commonly used to indicate response in studies of airway sensitivity and in evaluating experimental exposures of humans to inhaled toxicants. Although resistance can be measured during tidal breathing with esophageal catheters or oscillation methods, it is most commonly measured with plethysmography (Leith and Mead, 1974; Zarins and Clausen, 1982). The subject is seated within a body plethysmograph (a box with transducers that sense changes in pressure) and breathes with a panting pattern while flow at the mouth and pressure changes within the plethysmograph are measured. The airway is then occluded, and mouth pressure is measured as representative of alveolar pressure. The resulting data are used to calculate resistance and thoracic gas volume. Resistance can depend on volume, so it is often divided by volume and expressed as specific airway resistance (resistance per unit volume) or its reciprocal, specific airway conductance.

Current measurements of dynamic lung mechanics are useful clinical tools, but are unlikely to gain substantially broader use in population studies. The information obtained represents the integrated response of the entire lung; the lack of regional specificity and the lack of sensitivity due to intersubject variability reduce its utility as a marker of subtle effects. The greatest potential for development as a marker appears to lie in the use of oscillation methods because they are noninvasive and provide considerable information without requiring difficult procedures as described below.

Respiratory System Impedance (Oscillation Mechanics)

Measurement of "oscillation mechanics" is a means of evaluating the mechanical properties of the respiratory system. The technique provides a marker of response in the form of information on changes in the mechanical properties of the lung. The technique is rapid and and requires little cooperation from the subject—characteristics that make it suitable for epidemiologic studies. It provides indexes of lung compliance, as well as resistance; thus, it has potential for adding to the spectrum of information obtained in population studies. Tests of oscillation mechanics are in use for measuring the integrated compliance and resistance of an entire lung, but its potential as a marker is primarily in describing mechanical properties of specific regions of the respiratory system. The extent of the potential is uncertain and is the focus of current developmental work.

Oscillation mechanics was recently reviewed by Peslin and Fredberg (1986). The general approach is to superimpose an oscillating pressure signal on the airway during normal tidal breathing with a loudspeaker or pump. The frequency of oscillation is higher than the respiratory frequency of the subject, and the oscillating pressure and flow changes are small. The resulting pressure, volume, or flow perturbations in the air column are measured and used to calculate values of components of the mechanical impedance (resistance, compliance, and inertance) of the respiratory system.

The response of the respiratory system to an oscillating signal is determined by its impedance, which in turn is determined by its anatomic and mechanical properties. The overall response of the system represents an integration of the elastic, resistive, and inertial characteristics of each component of the system. In the simplest form of the assay, oscillation at a single frequency is used to measure dynamic compliance and airway resistance of the entire lung, without the need for a body plethysmograph or esophageal catheter. By manipulating the oscillating signal and analyzing the resulting response, one can theoretically extract information specific for different mechanical properties and for different anatomic structures.

Two approaches have been used for interpreting respiratory system impedance. One is the empirical association of changes in impedance with lung abnormalities (Kejeldgaard et al., 1976). The second is the estimation of specific impedance parameters by fitting impedance data to mathematical models of the respiratory system, which are based on mechanical or electric analogues (Jackson et al., 1984; Peslin et al., 1986). The latter approach should provide more descriptive information, if model parameters can be correlated with physiologic elements of the respiratory system.

Much of the effort in this field is directed toward development of improved models of respiratory system impedance. Previous work focused primarily on oscillating frequencies of 2-32 Hz, and it now appears that such data allow reliable extraction of only the integrated compliance, resistance, and inertance of the total respiratory system (Jackson et al., 1984). By extending the range of oscillating frequency, one can obtain statistically reliable estimates of additional parameters. For example, Jackson and Watson (1982) differentiated between central and peripheral resistance, compliance, and inertance by fitting oscillation data from rats to a six-parameter model. Their group has obtained similar results with other animal species, but has encountered difficulties in applying such models to data from humans. They hypothesize that models for humans need to account for shunting of flow in upper airways and for acoustic phenomena that occur in the relatively long airways of humans. No models have yet been shown to be satisfactory for clearly discriminating between mechanical properties of central and peripheral airways of human lungs.

It is not clear whether oscillation measurements will constitute improved tools for detecting and describing abnormalities of respiratory system mechanics due to environmental exposures. Considerable work remains to be done to develop appropriate models and to confirm associations among impedance changes, physiologic correlates, and alterations in respiratory system structure. That will require both application of the method to patients with known abnormalities of representative types and the study of animals with specific, experimentally induced abnormalities. Those lines of research are just now being pursued, and the utility of the approach is not likely to be fully known for a few years.

Measurements of oscillation mechanics with substantially improved descriptive value beyond that of tests currently in use would require specialized equipment. The oscillating system would consist of computer-generated signals fed to carefully calibrated loudspeakers or pumps. The frequency-response characteristics of the measurement system would have to be optimized. The data-reduction and model-fitting systems would be computer-based. Although the equipment would be specialized, it could probably be packaged into a mobile unit that could be operated by people with only modest training. Professional input would be required for cali-

bration, supervision of maintenance, and interpretation of results.

In summary, oscillation mechanics has potential for development into a useful marker of response. Its advantage lies in its ability to distinguish mechanical abnormalities on a site-specific anatomic basis. Its primary disadvantages are its dependence on an appropriate model for fitting data and the likelihood of substantial variation among individuals in regional mechanical properties of the respiratory system. General acceptance and widespread use will require substantial effort to demonstrate physiologic and clinical correlates, standardization of procedures and analysis, and packaging into measurement systems that are readily used.

Static-Quasistatic Lung Pressure—Volume Analysis

Lung compliance measured during breathing is usually lower than the actual compliance of lung tissue, because of the lack of time for tissue relaxation and because of differences in compliance among lung units. Measurement of static or quasistatic compliance avoids such frequency dependence by plotting transpulmonary pressure against lung volume during a single, slow exhalation. The current tests are assays that examine the elastic properties of lung tissue most directly and are the procedures of choice, if a specific index of lung elastic recoil is desired as a marker of response.

Standardized procedures for measuring lung compliance were recommended in a report from NIH (Macklem, 1974). Transpulmonary pressure is measured with an esophageal balloon catheter (Dawson, 1982). The subject inhales to total lung capacity and then exhales slowly while the exhaled volume and transpulmonary pressure are recorded. The test is termed quasistatic if exhalation is continuous, and static if exhalation is interrupted periodically to allow flow to cease and elastic forces to come to equilibrium. The elastic characteristics of the lung are expressed either by calculating compliance as the slope of some portion of the pressure—volume curve or by simply displaying the entire curve.

The lung pressure-volume curve shifts to the left (compliance increases) when lung elastic recoil is reduced (e.g., in emphysema) and shifts to the right (compliance decreases) when elastic recoil is increased (e.g., in fibrosis) (Macklem and Becklake, 1963). The curve represents the integrated elastic characteristics of the entire lung. It is not specific for the anatomic site of the change in recoil. Nor is it specific for the cause of the change in recoil. For example, fibrosis, inflammation, edema, and proliferative disorders could all cause similar shifts of the curve to the right (showing reduced compliance). Regardless, the test could be a useful marker of response in populations in which abnormal elastic recoil is a likely response.

Intrapulmonary Gas and Particle Distribution

Gas Distribution Properties— Single-Breath Gas Washout

Patterns of the washout of gases inhaled in a single breath have received considerable attention as indexes of small airway disease. Although the test can be performed by having the subject inhale a bolus of inert gas (bolus technique), the most common approach is to evaluate the washout of nitrogen from the lung after an inhalation of oxygen (resident-gas technique).

The single-breath nitrogen washout (SBNW) test was introduced in 1969 as a test of small-airway function (Anthonisen et al., 1969). The air-breathing subject exhales to residual volume, inhales a single breath of oxygen, and exhales again to residual volume. The nitrogen concentration of the expirate is plotted against its volume. The normal curve has a characteristic shape, first noted in 1949 (Fowler, 1949), in which the nitrogen is initially low (washout of dead-space oxygen), rises to a plateau that represents the nitrogen-oxygen distribution in the majority of the lung, and then increases again near the end of the exhalation. The

slope of the curve depends on the uniformity of gas distribution among ventilating units and is affected both by asymmetry of airway path lengths and by nonuniformity of compliance among ventilating units (Engel and Macklem, 1977). The slope increases as gas distribution becomes less uniform. The onset of the terminal nitrogen rise has been termed "closing volume" and is thought to indicate the lung volume at which airway closure begins (Engel et al., 1975). The phenomena responsible for determining the closing volume remain incompletely defined, but it is generally agreed that an increase in, if not onset of, airway closure is primarily responsible (Forkert et al., 1979).

It is interesting that Ernst et al. (1986) examined the relationship of closing volume and fluoride air pollution in children living near an aluminum smelter. In both sexes, there was a significant linear relationship between increased closing volume and the amount of fluoride found in urine samples from the children.

Since 1969, the SBNW test has been the focus of numerous physiologic studies and has been applied in several population studies. Standardized measurement procedures were disseminated by NIH in 1973 (Martin and Macklem, 1973), partly to facilitate multi-institutional collaborative studies funded by the National Heart and Lung Institute (NHLI). Methods for computerizing analysis of the curves have been published (Craven et al., 1976; Cramer and Miller, 1977). Equipment for performing the test is available commercially, and several equations have been developed for predicting normal values of SBNW parameters (Gold, 1982).

Although the SBNW test continues to be included in lists of tests sensitive to small-airway disease, its usefulness as a marker of responses to environmental exposures has not been clearly demonstrated. In 1973, a workshop on screening programs for early diagnosis of airway obstruction (NHLI, 1973) concluded that, "although closing volume and closing capacity are sensitive tests, they are probably of low specificity and moderate precision, and their validity as an early diagnostic test is unknown." The presumed usefulness of the test is founded largely on the finding in numerous studies that it can detect abnormalities in asymptomatic smokers, often in the absence of abnormalities in "conventional" lung function tests (McCarthy et al., 1972; Buist and Ross, 1973; Nemery et al., 1981; Teculescu et al., 1986).

Recent work has more directly demonstrated associations between SBNW abnormalities and small-airway pathologic conditions. Cosio et al. (1978) and Berend et al. (1981a,b) found significant correlations between abnormal values of washout slope and closing volume in human subjects and small-airway disease in excised lung tissue. The latter study demonstrated that closing volume was related more closely to small-airway inflammation than to lung elastic recoil. Petty et al. (1980) performed SBNW tests on excised human lungs and found that increased closing volume was associated with inflammation and squamous metaplasia in small airways. Those results further confirm and define the morphologic basis for SBNW abnormalities.

Incalzi et al. (1985) recently published regression equations of SBNW parameters with age, height, and lung volume for 234 normal subjects 20-80 years old with no history of smoking, occupational exposure to known pulmonary toxicants, or chronic respiratory illness. The authors concluded that the variability was too great for detection of subtle changes in population studies.

In summary, the utility of the SBNW test as a marker of responses to environmental exposures remains uncertain. The test reflects small-airway abnormalities, but its sensitivity and specificity are questionable.

Inhaled-Particle Distribution

The deposition of inhaled particles is a function of particle characteristics, airway geometry, and ventilation. The latter two can be altered by exposure to pollutants or by disease, so it follows that aerosols can be used as markers of exposure or response to environmental agents. Although the specific equipment required for tests of aerosol distribution

is not generally available in the standard pulmonary function setting, the technology is neither new nor very complicated. Establishment of guidelines for their use, which do not now exist, could result in uniform application in the future.

Aerosol particles can be used to assess pulmonary structure and function, because they can trace the convective motion of air in the lungs and their deposition is related to the dimensions of the airways through which they pass. Three techniques can be used to obtain information from inhaled particles; they allow assessment of airway sizes and inhomogeneities of ventilation and gas mixing. The use of aerosols to assess mechanical clearance from the respiratory tract is discussed later.

Assessment of gas mixing. Intrapulmonary mechanical mixing of gases can be assessed by injecting an aerosol during the entire tidal-volume inhalation or in a pulse during a portion of this inhalation and then examining the particle concentration recovered in exhaled air. The procedure requires the use of particles with a minimal probability of deposition—approximately 0.3-0.5 μm—so that their loss from the inhaled air occurs largely because of nondiffusive gas mixing in the lungs, i.e., bulk transfer from tidal to reserve air. By separating mixing due to molecular diffusion from mechanical mixing due to airflow, one can estimate the role of molecular diffusion in ventilation (Altshuler et al., 1959). Although the procedure provides some assessment of bulk transfer, conclusions as to the sites at which this occurs await the further development and use of models of aerosol dynamics and gas transport in the lungs that take into account the effects of geometric complexities on lung ventilation (e.g., Engel, 1983).

The distribution of exhaled aerosol in normal people shows a fair degree of intersubject variability; nevertheless, general profiles are reproducible within groups of subjects, and exhaled-aerosol measurements have been used to assess airway abnormalities, in which case the shape of the aerosol exhalation curve is different from that in normal subjects. In airflow obstruction, for example, the shape of the curve is different because the recovery of particles is decreased, owing to an increased rate of particle deposition and a change in mixing characteristics. A correlation has been found between the percentage of aerosol recovered and the predicted percentage change in FEV_1 over a wide range of degrees of airway obstruction (Muir, 1970). A difference in aerosol recovery (i.e., a decrease) has also been demonstrated in coal miners with various forms of pneumoconiosis (Hankinson et al., 1979).

Aerosol probe procedures. The aerosol probe technique allows inferences concerning small-airway (< 1 mm) and alveolar dimensions; it is of particular use for assessing changes in diameter, such as those associated with obstructive lung disease. Conducting-airway obstruction can also be detected with conventional pulmonary function tests, but the latter might be less sensitive than the probe procedures in detecting early changes. However, alveolar size, which is important in assessing emphysema progression, can be estimated in vivo only with particle probes.

The aerosol probe procedure is a modification of the aerosol mixing technique discussed above, in that a period of breath-holding is generally imposed between aerosol inhalation and exhalation. It is based on the concept that the amount of deposition of inert, nonhydroscopic monodisperse particles with a known, but low, rate of sedimentation during the breath-holding period depends on the settling distance required for the particles before they come into contact with an airway wall; this distance is a reflection of the overall dimensions of the airspaces in which the aerosol is found. Particles that are deposited and are removed from the air will not be recovered during exhalation. In reality, and as should be evident from the discussion of the mixing technique above, not all the inhaled aerosol will be recovered even if there is no breath-holding period between inhalation and exhalation. In practice, the aerosol probe procedure requires determination of the extra loss of particles due to gravi-

tational settlement. The ideal particle size range used in the procedure is 1-1.5 μm MMAD (Gebhart et al., 1981); however, the increased impaction deposition in people with obstructed airways might require some size adjustment.

The work of Palmes et al. (1967) established a basic method for estimating the effective dimensions of the respiratory airspaces with monodisperse aerosols. An aerosol is inhaled; the breath is then held at close to total lung capacity (TLC), with different breaths held for periods of 0-30 seconds; and a volume equal to twice the inhaled volume is then exhaled, to ensure recovery of all remaining airborne particles. The persistence of the aerosol—i.e., the probability of its remaining suspended in air and thus being exhaled—decreases exponentially as a function of breath-holding time. Aerosol inhalation was completed near TLC, so most of the aerosol mass and most of its deposition were assumed to be in the region of the respiratory bronchioles and alveolar ducts; the contribution of anatomic dead-space volume was considered to be small, and its influence on the shape of the aerosol recovery curve could be ignored. The logarithm of percent of aerosol recovery is plotted against breath-holding time; this results in a curve whose slope (or slopes) is related to the average size of the airways within which residual aerosol remained before exhalation. Results are usually expressed in terms of the half-time of aerosol persistence in the lung.

Subjects differ substantially, but the aerosol probe procedure is sensitive to changes in airway dimensions and does yield reproducible results in normal subjects repeatedly tested (Lapp et al., 1975); in addition, the variability in airway size measured in healthy people was found to be quite similar to that measured in fixed lungs obtained from accident victims. However, the wide intersubject variability in normal people is a negative feature of the test if applied to people with lung abnormalities; e.g., in obstructive disease, the results can be equivocal (Palmes et al., 1971, 1973). In some cases, increased half-time of aerosol persistence indicates enlarged airspaces; but results in patients with diagnosed emphysema can be comparable with those in healthy people. Inasmuch as the aerosol has access only to airspaces to which it is delivered by convective airflow, it can penetrate to either predominantly normal or diseased lung tissue, depending on the site of airflow obstruction (if any). Thus, half-times observed in patients are more variable than those observed in healthy subjects. In addition, the amount of aerosol at zero time of breath-holding was generally lower in patients than in normal persons; that indicates increased deposition during the dynamic phase of the breath-holding maneuver, probably due to obstruction and the resulting narrowing of airways.

Another group examined with the aerosol probe were coal miners with pneumoconiosis (Hankinson et al., 1979). There was some correlation between disease type, aerosol persistence, and calculated airway dimensions, but a lack of correlation between persistence and recovery suggested either that the mechanisms that cause changes in those two phenomena are different or that the changes occur at different sites in the respiratory tract. The latter possibility is of concern, because the results of the tests are a reflection of airway dimensions at various depths in the lungs.

A slight modification of the aerosol probe technique might be used to examine particular regions (depths) in the lungs. The modification, known as the bolus probe technique, involves either inhalation of an aerosol bolus followed by a preset (but variable) volume of particle-free air (Palmes et al., 1973) or inhalation of the bolus at various stages of inhalation rather than at a particular fixed stage (Heyder, 1983). When this procedure is used, inhalation of equivalent tidal volumes leads to decreases in measured effective airway diameter as lung volumes decrease. In addition, the measured persistence and thus the actual average airway diameter that is measured depend heavily on the inhalation volume containing the aerosol, in that, the more deeply a bolus is inhaled, the greater is the dispersion of the bolus later exhaled and the smaller

are the airways being "probed." The airway dimensions calculated from aerosol recovery curves for different depths of inhalation have been found to agree well with what would be expected, on the basis of comparisons with both morphometric models of the human lung and measured airways in fixed lungs (Gebhart et al., 1981; Heyder 1983; Nikiforov and Schlesinger, 1985).

In an aerosol rebreathing procedure described by Kim et al. (1983), subjects breathe $1-\mu m$ particles 30 times/minute from a volume held at 500 ml. That results in magnification of the differences in particle recovery after a single breath. The test is able to screen reproducibly for airway constriction and might also indicate the extent of such concentration; variability in both normal subjects and those with chronic obstructive airway disease was 5-10%. The ability of the test to detect changes early in pathogenesis is not yet known (C. S. Kim et al., 1985).

A variation of the single-breath aerosol bolus technique that is sensitive in detecting early airway changes has been described (McCawley and Lippmann, 1984; McCawley, 1987). It involves precise control of volume and flow during introduction of a bolus of monodisperse ($0.5-\mu m$ MMAD) particles. With a dispersion index, it was possible to differentiate between healthy smokers and nonsmokers; thus, the procedure seems capable of detecting the early changes known to occur in the small airways of smokers. There were found to be no differences in tests of forced expiration (e.g., FEV_1 and FVC) between smokers and nonsmokers, and the coefficient of variation of the dispersion parameter (approximately 17% in nonsmokers and 36% in smokers) was less than that of the pulmonary mechanics tests. The procedure optimizes the protocol for rapid screening and is useful in epidemiologic studies that attempt to assess airway obstruction.

Regardless of the specific probe procedure used, there are always differences between the simple theoretical assumptions of the models of aerosol deposition and actual complex situations. Direct anatomic interpretations should be made with caution, and one must bear in mind effects of a number of factors, e.g., axial dispersion of the aerosol bolus, nonuniformity of the aerosol concentrations in cross-section at the end of inhalation, the relative shortness of the airways (which become increasingly short with greater depth in the lung), the irregularity of branching, and the regional distribution of airflow in the lungs.

Radioimaging Techniques. Particles with MMAD of 1 μm or more are subject to deposition by impaction and thus might provide information on obstruction in small airways before it is measurable with conventional pulmonary mechanics tests. In all cases, the pattern of deposition depends on the distribution of ventilation in the lungs; in obstructive disease, deposition is also influenced by the sites of obstruction, and there is no deposition in regions where there is no ventilation. Radioimaging procedures involve the external detection of deposited particles labeled with a radioactive tag. The types of detectors used are the same as those used in measuring particle clearance and are discussed in Chapter 2.

In normal people, the radiolabeled-aerosol image appears to indicate a fairly even distribution of radioactivity, especially if particles 1-4 μm in diameter are inhaled at flow rates that occur at rest. However, obstructive disease or obstruction is indicated by the presence of areas of concentrated deposition and areas of minimal or no deposition. Much of the former occurs in the central bronchi, and the latter in peripheral areas.

Early studies (Ramanna et al., 1975; Taplin et al., 1977) indicated qualitative differences in the pattern of deposition between normal people and people with clinical indications of airway obstruction; the differences were ascribed to differences in regional ventilation. The quantitation of particle deposition to assess obstruction is generally performed in two ways. One procedure is to use a measure of retention at an appropriate time after exposure to the tracer particles as an index of regional deposition, i.e., tracheobronchial versus alveolar. Generally, retention at 24 hours is the measure used. That is not necessarily accurate, in that changes in particle clearance rates

might alter amounts present and lead to false conclusions as to regional deposition. In addition, 24-hour retention varies widely in normal people and even more in smokers and people with lung disease.

Another technique makes use of gamma-camera imaging to differentiate deposition in central (hilar) and peripheral regions of the lungs (Dolovich et al., 1976; Emmett et al., 1984). In the simplest procedure, which uses a calibration with an 81-mKr ventilation scan, it is possible to obtain the ratio of outer to inner zone radioactivity of the tracer particles and to describe a "penetration" index; the index relates the penetration of radiolabeled aerosol into peripheral airways to the degree of ventilation of these airways. Such procedures do show differences between healthy, nonsmoking subjects and asymptomatic smokers; thus, they can detect changes often not seen with conventional pulmonary function tests. Penetration index has also been found to correlate fairly well with pulmonary function measures, such as FEV_1, in people with diagnosed chronic obstructive disease (Agnew et al., 1981; Agnew, 1984). Unfortunately, the tests appear to yield a high number of false-positives, even though they are very sensitive.

The above procedures provide no information on the precise anatomic location of deposited particles. An approach to assessing gamma-camera images to allow better definition of sites of deposition involves development of concentric zones around the hilar region, each of which constitutes a different percentage (e.g., 25%, 50%, and 75%) of the entire lung field (Foster et al., 1985). The innermost 50% is considered as the central fraction.

Differences in aerosol penetration sometimes observed between smokers and nonsmokers have resulted in the suggestion that radioimaging procedures can be used to produce markers of developing or early disease. They also seem better as sensitive indicators of obstruction in large airways than in small airways (Chopra et al., 1979). However, there is an inherent problem in the techniques: they require the use of radioactive aerosols on a routine screening basis. Thus, such procedures should not be used if other, more conventional tests are available for screening.

Alveolar–Capillary Gas Transfer

Diffusing Capacity for Carbon Monoxide

The diffusing capacity for carbon monoxide (DCCO), sometimes called "transfer factor," is considered the most sensitive measure of alveolar-capillary gas-exchange efficiency performed with the subject at rest and is recommended by the American Thoracic Society (1986) as one of the two primary measures for evaluating impairment of respiratory function (spirometry is the other). DLCO is the rate of uptake of a low concentration of inhaled CO, normalized by the alveolar-capillary difference in the partial pressure of CO. CO is used as the indicator gas, because of its high affinity for hemoglobin, which, at low concentrations, obviates the sampling of blood to measure the pressure difference.

There are several methods for measuring DLCO and multiple variations of each method; however, the single-breath technique (Ogilvie et al., 1957) is preferred. The measurement technique and methodologic factors that influence the results were described in detail in a recommendation for standardization published by the American Thoracic Society (1987). The subject inhales a large breath of a mixture of CO and an inert gas in air, holds the breath for a few seconds, and exhales. A sample of "alveolar" gas is taken late in exhalation, and concentrations of CO and the inert gas are measured. The inert gas allows the lung volume at end inspiration (termed "alveolar volume," although airway volume is also included) to be calculated. DLCO is volume-dependent, so the DLCO:alveolar volume ratio helps in determining the contribution of reduction in lung volume to reduction in DLCO.

As with most measures of respiratory function, DLCO is influenced by several factors; thus, the test has little specificity. DLCO is influenced by membrane factors, such as thickness and surface area; by capillary blood volume; and by the rate of combination of CO with hemoglobin. The latter factor makes the test somewhat sensitive to hemoglobin concentration and to altitude. Methods for correcting for hemoglobin concentration were included in the recommendations of the American Thoracic Society (1987). Ayers et al. (1975) reviewed some of the physiologic factors that influence DLCO and a scheme for differential diagnosis.

Of the tests performed at rest, measurement of DLCO has been shown to be the most sensitive to some abnormalities, such as radiation-induced pneumonitis in humans and animals (Mauderly et al., 1980). It was the first measure shown to change in a progressive disease. Its inherent sensitivity is limited, however, by the presence of considerable variation among individuals. Coefficients of variation among individuals are typically about 4% for normal persons and 7% for subjects with severe obstructive disease (American Thoracic Society, 1987). It has been recommended that, for a subject to be considered "mildly impaired," the measured value be below 79% of predicted (American Thoracic Society, 1986). Generally accepted formulas for predicting normal values are available (Crapo and Morris, 1981).

Automated equipment for measuring DLCO is available commercially and is in general clinical use. The test requires sufficient subject cooperation to perform the single-breath maneuver, but is rapid and readily adapted to mobile facilities. The trace amount of CO used in the test is innocuous. The test could be used in large population studies; however, because of variability, it would have questionable utility in such a setting. It would be more useful in studies of more selected populations, such as occupational cohorts. The test is well developed; recent progress has been largely in standardization of technique and development of commercially available equipment for reproducible performance.

Gas Exchange During Exercise

A major problem in using respiratory function tests at rest to detect subtle abnormalities is the large reserve functional capacity of the lung. Exercise testing provides an examination of the performance of both the respiratory and cardiovascular systems under metabolic conditions that increase the demand on them. Although exercise testing requires cooperation from the subject and takes longer than the functional assays described above, it need not be invasive (i.e., blood samples are not necessarily required), and it can be performed in the field. If blood samples are obtained, the approach should provide greater sensitivity to subtle gas-exchange abnormalities than any of the tests performed at rest. Excellent reviews of exercise testing have recently been published. Wasserman et al. (1987) presented a thorough review of theoretical and practical aspects, including case reports; and Hansen (1982) presented a concise review, including procedures, equations, and sources of error. Only selected facets of the approach are discussed here.

Gas exchange between the cells and the environment requires the effective interaction of the lung and chest "bellows" (thoracic wall and diaphragm), a heart capable of pumping sufficient blood, a vascular system that can selectively distribute flow to match requirements, and respiratory control mechanisms capable of regulating blood-gas tensions and pH. In a healthy person, the responses of those systems to exercise are predictable and depend only on the rate of work performed and the fitness of the subject (Wasserman et al., 1981). The usefulness of exercise testing is based on the ability to measure deviations from predicted physiologic behavior.

There are several variant protocols for exercise testing, but all have the goal of using work by large muscle groups to increase metabolic demand. Both treadmill exercise and cycle ergometer exercise are used, but the cycle is usually preferred. The work expended on a cycle is more readily calibrated than that on a treadmill. Variability during treadmill exercise associated with body build, pattern of leg movement, and holding onto side rails is minimized with a cycle, and there is less risk of slipping. Both prolonged, steady-state exercise and non-steady-state incremental exercise are used. The latter might be preferable, because it can be matched to the subject's ability, and the various factors that affect performance can be distinguished. The basic measurements include electrocardiography, determination of ventilatory flows and volumes, and measurement of oxygen and carbon dioxide concentrations at the mouth. Manual collection of exhaled gas is usually obviated by computer integration of breath-by-breath volumes and gas concentrations; however, manual collection is possible. Measurements of arterial pressure, blood gases, alveolar-arterial gas tension differences, and acid-base status contribute substantially to the information gained from the test, but require arterial puncture. Much information can be gained, however, from numerous other measurements that are.

Characteristics that can be measured during exercise and their interpretive usefulness have been listed and discussed at length (Wasserman et al., 1987). The anaerobic threshold and relationships between heart rate and oxygen uptake appear to have good overall sensitivity to gas exchange and cardiovascular inefficiencies. The anaerobic threshold is the highest oxygen uptake that can be sustained without metabolic acidosis, and it is lowered by any factor that limits oxygen flow to exercising muscles. Although blood samples are required for precise definition of the anaerobic threshold, it can be satisfactorily estimated noninvasively from a plot of carbon dioxide output versus oxygen uptake over a range of work magnitudes. The slope of the ratio of heart rate to oxygen uptake, plotted over a range of work magnitudes, is sensitive to both airway and heart abnormalities and can sometimes be used to distinguish between the two. The "oxygen pulse," oxygen uptake per heartbeat, equals the product of stroke volume and the arteriovenous oxygen difference and is also sensitive to anemia and carboxyhemoglobin.

Several additional characteristics can be measured noninvasively, including maximal exercise ventilation, ratio of dead space to tidal volume, ventilatory equivalents for oxygen and carbon dioxide, respiratory-exchange ratio (ratio of CO_2 output to O_2 input), expiratory flow patterns (examined as are those obtained during forced exhalation), and ratio of tidal volume to inspiratory capacity (a reflection of lung stiffness). Each of those has a relatively specific sensitivity for particular physiologic abnormalities.

The addition of arterial-blood sampling extends the usefulness of exercise testing greatly. Not only are blood-gas tensions and acid-base status of interest, but the alveolar-arterial differences in oxygen and carbon dioxide tensions during exercise are probably the most sensitive of all tests of pulmonary gas exchange at the alveolar level.

Exercise testing is being applied to clinical subjects in numerous laboratories. Application in population studies (discussed by Wasserman, 1985) has been limited. Sue et al. (1987) compared the sensitivities of exercise testing and resting single-breath CO diffusing capacity (DLCO) to detect abnormalities among 276 current or former shipyard workers. The criterion for abnormal DLCO was set at 70% of predicted, and several prediction equations were evaluated. Only 16 subjects had an abnormal DLCO, and exercise gas exchange was abnormal in all but two of them. In contrast, resting DLCO was abnormal in only 14 of 96 men with abnormal exercise gas exchange. Those results suggest that exercise testing can greatly improve the detection of subtle gas-exchange abnormalities, in addition to providing considerable discrimination among the factors that cause the abnormalities.

In summary, exercise testing appears to have one of the strongest potentials among respiratory function assays as a sensitive marker of pulmonary injury associated with environmental exposures. The degree of interaction required from the subject and the extended usefulness of the assay if blood samples are included tend to make it more suitable for studies of limited populations in stationary facilities under the direct supervision of experienced researchers. Equipment for performing exercise tests in a standard-ized manner is now becoming available commercially. It is practical to consider use of at least some aspects of exercise testing in mobile facilities in the field. Exercise testing would have particular usefulness in studies in which gas-exchange or cardiovascular abnormalities were of primary interest.

AIRWAY HYPERREACTIVITY

Although airway provocation testing dates to the late 1940s, it did not emerge as a clinical diagnostic and research tool until the early 1970s. Nonspecific airway hyperreactivity is defined as an exaggerated bronchoconstrictor response to a variety of chemical, physical, and pharmacologic stimuli. There is now nearly a consensus that nonspecific airway hyperreactivity is a characteristic shared by virtually all asthmatics (Boushey et al., 1980); that is, the asthmatic develops bronchoconstriction after inhaling a lower concentration of a provoking agent than is needed to cause a similar degree of change in airway tone in a healthy subject. Thus, in one sense, airway hyperreactivity serves as a marker of asthma. Its mere presence alone does not define asthma, however, inasmuch as increased bronchial reactivity can occur in otherwise healthy people. Furthermore, airway hyperreactivity has been observed in healthy people after viral upper respiratory infections (Little et al., 1978), after acute inhalation of ozone (Golden et al., 1978) or sulfuric acid aerosols (Utell et al., 1984), or after inhalation of irritants in the workplace (Brooks et al., 1985). Whether such transient increases in airway hyperreactivity serve as markers of long-term respiratory sequelae is unknown, but they might well offer a clue to the pathogenesis of chronic respiratory diseases.

Nonspecific airway hyperreactivity presumably reflects or mimics events in naturally occurring asthma. Thus, it should not be surprising that many of the proposed mechanisms of asthma have also been linked with the exaggerated contraction of smooth muscle that characterizes airway hyperreactivity. However, it should be noted that airway hyperreactivity does not correlate with histamine reactivity of bronchial smooth muscle (Roberts et al., 1985). Possible mechanisms include alterations in airway geometry, disorders of autonomic regulation of smooth muscle, structural alteration in airway smooth muscle, increased accessibility of stimuli to the muscle, and release of locally acting mediators of inflammation (Boushey et al., 1980; Sheppard, 1986b). Present evidence suggests that both neural and nonneural mechanisms contribute to airway hyperreactivity. For example, abnormally small baseline airway caliber increases reactivity, but cannot explain the hyperreactivity seen in asthmatics in remission or in persons with respiratory viral infections. Autonomic regulation of the airways seems an attractive explanation in studies in which responses can be abolished by directly interfering with nerve transmission or inhibited by pretreating with blocking agents (Nadel and Barnes, 1984). Epithelial injury from viral infections or inhalation of pollutants might result in more direct exposure of nerve endings to an agent that provokes bronchial response, but evidence of such injury is often not present. Recently, the development of acute inflammatory response has been linked to the pathogenesis of airway hyperreactivity, especially after ozone inhalation (O'Byrne, 1986). Several products of arachidonic acid metabolism released during the inflammatory response (Sheppard, 1986a) or by immunologic activation of human lung mast cells (Schulman, 1986) have been targeted as possible mediators. If inflammation and mediator release are linked with the acute development of airway hyperreactivity, then recurrent episodes would have a greater likelihood of producing chronic airway effects.

Methods of Assessment

In the laboratory, airway reactivity testing is divided into two general categories, depending on the choice of nonspecific versus specific agents. In both, the increased bronchoconstrictor response is assessed with pulmonary function tests.

Nonspecific stimuli include pharmacologic agents, such as methacholine, carbachol, and histamine; exercise; hyperpnea with cold or dry air; and inhalation of hypertonic or hypotonic aerosols. Although pharmacologic challenge is used most often in the clinical laboratory, it is less suitable for population studies, especially those involving children. Cold-air challenge with hyperventilation has been used effectively, and response is generally correlated closely with methacholine responses. Challenges with specific agents, such as common antigens, such chemicals as isocyanates, and such organic materials as plicatic acid (from western red cedar) attempt to identify specific sensitizing agents. Those approaches can be particularly powerful in incriminating occupational chemicals and confirming the diagnosis of occupation-related airway disease, but might provoke immediate and/or late pulmonary responses that do not resolve spontaneously. Even with specific agents, the interpretation of responses can be difficult and confounded by a variety of factors, such as dose and irritant effects (McKay, 1986).

Inhalation challenge testing with cholinergic agonists and histamine has varied considerably among laboratories, and efforts to standardize it have been advocated (Hargreave and Woolcock, 1985). Standardized approaches could minimize variability in aerosol generation and inhalation, in methods of measuring response and expressing results, in preparation and handling of test solutions, and in concomitant use of medications that affect bronchoconstrictor response. Aerosols are usually generated and inhaled with two techniques: continuous generation of an aerosol inhaled during tidal breathing and generation of a puff of aerosol with a dose-measuring device or hand-held nebulizer and inhalation of the puff during a single deep breath. Aerosol particles range from submicrometer size to several micrometers and from nearly monodisperse to widely heterodisperse. Such variables affect lung deposition and influence airway reactivity (Dolovich, 1985).

The bronchoconstrictor response is

commonly measured with two methods—on the basis of airway resistance and maximal expiratory flow (FEV_1). Both have shortcomings. The plethysmographic measurement of airway resistance includes resistance of the larynx, so an increase in resistance could reflect laryngeal narrowing or smooth muscle contraction. One potentially confounding problem with expiratory maneuvers is that the test itself might alter the characteristics being tested. The deep inhalation that precedes measurements of expiratory flow causes transient bronchodilation and lessens the bronchoconstriction in a normal subject, whereas it might increase bronchoconstriction in an asthmatic. Nevertheless, the bronchoconstrictor response is usually measured as the change in FEV_1. Increasing doses of an aerosol are inhaled to construct a dose-response curve, and the results are expressed as the provocation concentration (pc) necessary to produce a decrease in FEV_1 of 20% (PC_{20}). The PC_{20} is obtained from the log-dose-response curve by linear interpolation of the last two points; the lower the PC_{20}, the greater the reactivity. PC_{20} has been found to correlate closely with methacholine and histamine concentrations in the aerosol (Hargreave et al., 1983).

An alternative approach is provocation testing with physical stimuli, such as exercise or isocapnic hyperventilation with cold air. Those are naturally occurring stimuli and their use obviates the inhalation of pharmacologic agents. They can cause bronchoconstriction by airway cooling or airway drying, as well as through the endogenous release of mediators in the airway (Barnes and Brown, 1981). The degree of airway reactivity to methacholine or histamine tends to be correlated with reactivity to exercise or isocapnic hyperventilation. That suggests that the reactivity to chemical mediators is an important determinant of the response to exercise and hyperventilation. When assessing airway reactivity with exercise or isocapnic hyperventilation, one must control such variables as magnitude of ventilation, duration of exercise or hyperventilation, workload, and inhaled air temperature and water content (Hargreave and Woolcock, 1985). Although the provocation tests effectively identify hyperresponders, they provide less quantitative dose-response information. Otherwise there is little justification for concluding that one method is superior to another for assessing airway hyperreactivity; the choice depends on the experience of the investigator and the laboratory, although pharmacologic challenge might have the advantages of requiring less expensive equipment and being technically simpler.

Nonspecific airway hyperreactivity testing has proved to be highly useful for assessing airway responses to low concentrations of environmental air pollutants. Even after the return to baseline lung function on removal from acute nitrogen dioxide (Bauer et al., 1986) or sulfuric acid aerosol (Utell et al., 1984) exposure, asthmatics demonstrated increased airway reactivity to cold air and hyperventilation or to carbachol aerosols, respectively. Likewise, after removal of healthy volunteers from ozone environments, inhalation of methacholine or histamine increased bronchoconstriction (Golden et al., 1978).

In the assessment of asthma induced by occupational agents, airway reactivity testing with nonspecific and specific agents serves a diagnostic function. Chan-Yeung and Lam (1986) recently published a comprehensive review on the subject of occupational asthma and the role of airway reactivity testing. Nonspecific airway hyperreactivity occurs in most workers with occupationally induced asthma, despite the absence of predisposing factors, such as atopy. Furthermore, Lam and coworkers (1983) found a good correlation between the degree of nonspecific bronchial hyperreactivity and severity of response to the provoking agent plicatic acid in workers with red cedar asthma. Measurements of hyperreactivity also assist in providing objective evidence of sensitization (Chan-Yeung and Lam, 1986). The demonstration of an increase in bronchial reactivity on returning to the workplace and a decrease away from work, with appropriate changes in lung function, establishes the causal relationship between symptoms and the work environment. To pinpoint the etiologic agent in the workplace responsible for asthma, specific challenge is necessary. Chan-Yeung and Lam (1986) emphasized that such testing can be dangerous and should be performed only by experienced persons in a hospital setting for the following conditions: studying previously unrecognized occupational asthma, determining the precise etiologic agent in a complex industrial environment, and confirming a diagnosis for medicolegal purposes. Detailed guidelines and testing procedures have been developed and published (Pepys and Hutchcroft, 1975; McKay, 1986).

In summary, increased airway hyperreactivity is a hallmark of clinical asthma and occupationally induced asthma. In patients with asthma, the greater the reactivity, the greater the likelihood of symptoms. Despite variability in testing methods, the tests have proved reproducible, relatively simple to perform, and closely correlated with each other.

Distribution of Airway Hyperreactivity in the Population

There is a broad distribution of airway reactivity to cholinergic agents and histamine in the general population, with asthmatics among the most responsive group. However, increased airway hyperreactivity occurs in other pulmonary conditions, including cystic fibrosis, chronic bronchitis, and sarcoidosis. It is found in about 4% of people who have never had symptoms, in about 20% of patients with isolated chronic cough, and in about 10% of patients with rhinitis without chest symptoms (Hargreave et al., 1985). Several population studies have concluded that bronchial hyperreactivity occurs in a unimodal, rather than bimodal, distribution. For instance, in a distribution of bronchial reactivity, nonasthmatics could be placed at one end of the curve and asthmatics at the other, most reactive end. Subjects with hay fever or allergic rhinitis are often intermediate in nonspecific airway reactivity.

Few epidemiologic studies have examined the distribution of reactivity in unse-lected populations. In a study of 300 relatively young, randomly selected college students, Cockroft and colleagues (1983) studied nonspecific reactivity to inhaled histamine. Asthmatics responded to a concentration of histamine of 4 mg/ml or less. When responders were defined by such criteria, 20% of the nonasthmatic group demonstrated equivalent reactivity in the absence of symptoms. In a larger, population-based study, Weiss and colleagues (1984) assessed nonspecific bronchial reactivity with eucapnic hyperpnea in response to subfreezing air in 134 adults and 213 children. Children and young adults were significantly more likely than older subjects to be responders. Nearly 20% of asymptomatic children had increased airway hyperreactivity. Of current asthmatics, 92% responded to the cold-air challenge. One can conclude that the prevalence of increased reactivity in the population-based studies exceeds the prevalence of asthma in the general population. The observation that asymptomatic airway hyperreactivity occurs in many children and young people presents an opportunity to examine hyperreactivity as a marker of susceptibility to disease. Followup of such populations should determine whether bronchial hyperreactivity plays a role in the development of progressive lung disease; if it does, it would serve as a key risk factor (i.e., a marker of susceptibility to) for lung disease.

Airway Hyperreactivity as a Marker

Population-based studies that examine the effect of bronchial hyperreactivity on the rate of decline of pulmonary function in asymptomatic subjects are prerequisites to understanding airway reactivity as a risk factor. Several lines of evidence suggest that airway hyperreactivity is important in the development of chronic lung disease. Barter and Campbell (1976) demonstrated that subjects who were hyperreactive to methacholine had a more rapid decline in FEV_1 than did nonreactive subjects. Despite concerns that the two groups were not ideally matched and the need for further study, the findings are impressive.

Britt and colleagues (1980) reported that almost half a group of subjects who were first-degree relatives of patients with chronic obstructive lung disease showed increased reactivity to methacholine. In those with hyperreactive airways, a markedly accelerated loss of lung function was observed (150 ml/year), compared with normal loss of lung function in the nonreactors (30 ml/year). Thus, airway hyperreactivity might constitute a marker useful for detecting risk of accelerated loss of lung function. Efforts are needed to determine whether transient but recurrent episodes of airway hyperreactivity, such as follow pollutant exposure or viral respiratory tract infection, also serve as a risk factor for progressive lung disease.

Similar issues emerge in examining the relationship between airway hyperreactivity and occupational asthma. Although most patients with occupational asthma demonstrate nonspecific bronchial reactivity, it is unclear whether it is the result of exposure or a predisposing factor. However, most evidence points to airway hyperreactivity in this setting as a marker of exposure, in that reactivity often wanes with removal from the environment, it can recur with re-exposure, and it can persist for long periods in previously asymptomatic persons exposed to high concentrations of irritating gases (Brooks et al., 1985). Studies incorporating measurements of airway hyperreactivity in pre-employment examinations and after workplace exposure should help to resolve the issue of whether increased hyperreactivity is a marker of exposure or injury.

A large, multicenter study sponsored by the National Heart, Lung and Blood Institute is examining whether airway hyperreactivity affects (or predicts the rate of) loss of lung function in those at risk of chronic obstructive lung disease. If that is the case, airway hyperreactivity will prove to be a powerful marker for assessing susceptibility to environmental agents.

CLEARANCE OF PARTICLES FROM THE RESPIRATORY TRACT

Components of clearance systems are commonly the initial points of contact with inhaled pollutants; clearance function might therefore be useful as a marker of response or a predictor of respiratory tract disease. This discussion focuses on techniques that can be used to assess clearance in vivo; various elements of clearance systems can be examined with methods that use isolated organs, tissues, or cells. The latter could be important in attempts to evaluate individual components of clearance systems to explain mechanisms of impairment, but separation of mechanisms has generally not been possible, and the cause of clearance impairment cannot always be determined.

The broad mechanisms of clearance in conducting and respiratory airways are similar in humans and most other mammals. Interspecies differences do occur, for example, in secretory cell structure and distribution, macrophage function, relative roles of clearance mechanisms, and rates and efficiencies of clearance from various regions. Although such differences might be reflected in quantitative differences in response to inhaled agents, responses to numerous toxicants are qualitatively similar between humans and many experimental animals Phalen et al., 1984; Lippmann and Schlesinger, 1984). Thus, although there are anatomic and physiologic differences, they do not preclude the use of clearance as a marker of exposure or response to inhaled toxicants.

A valid marker should meet appropriate basic criteria. It should be reproducible (with minimal variability), it should be fairly sensitive to change after exposure to appropriate agents, and it should have a relationship to respiratory tract disease. Clearance rates from conducting and respiratory airways are well-defined functional characteristics of a person. However, interindividual variability is high. It is difficult to develop a library of values that are considered normal, for comparison with those obtained after pollutant exposure or in disease states.

Nevertheless, the current database does allow characterization of a person's clearance as within or outside a normal range (Agnew et al., 1986). Measured clearance rates depend heavily on the technique used to assess them, so some standardization of clearance measurement procedures is needed, if this function is to be used as a reproducible marker.

In both humans and experimental animals, bronchial mucociliary clearance alteration has been shown to be a sensitive physiologic indicator of response to exposure to many pollutants—such as nitrogen dioxide, ozone, and sulfuric acid—and occur after acute exposure to low concentrations of pollutants and early in chronic exposures (Wolff, 1986; Schlesinger and Driscoll, 1988). Studies with experimental animals have shown that alteration in rate of clearance from the respiratory region can be a sensitive indicator of exposure. Although transient alterations in clearance after acute exposure to inhaled irritants might be adaptive (helping to maintain organ homeostasis), such changes are more likely to be pathophysiologic responses of the airways and, although temporary, might foreshadow permanent alterations or progressive changes that would follow continued exposures (Schlesinger et al., 1983; Gearhart and Schlesinger, 1988).

Interpretation of mucociliary clearance alterations in terms of potential health problems is speculative. Dysfunction of mucus transport is probably involved in the pathogenesis of some acute and chronic respiratory diseases. An absence of mucociliary function in some people is directly responsible for the early development of recurrent respiratory tract infections and, eventually, chronic bronchitis and bronchiectasis (Wanner, 1980; Rossman et al., 1984). Partial impairment of the mucociliary system can also increase the risk of lung disease. In this regard, the rate of mucociliary clearance might affect the development of infectious disease. The upper respiratory tract is continuously exposed to potentially pathogenic organisms, and both these and organisms that reach the lower conducting airways are confronted with the mucociliary barrier.

The rate of penetration of mucus by disease vectors to the underlying cells relative to the rate of mucociliary transport out of the respiratory tract could determine the capacity of inhaled pathogens to initiate disease (Proctor, 1979; Niederman et al., 1983). There is a predisposition to respiratory infection in conditions (e.g., chronic bronchitis) characterized by retarded clearance from conducting airways (Cumming and Semple, 1980). In addition, destruction of the functional integrity of the ciliated epithelium results in impaired defense against bacteria (Laurenzi and Guarneri, 1966), and impaired transport has been observed in association with viral respiratory infections (Bang and Foard, 1964; Lourenco et al., 1971a).

Retardation of mucociliary clearance might also be a factor in the genesis of bronchial cancer, by increasing the residence time of carcinogens at initial deposition sites or those being carried through the lungs on the mucociliary escalator. For example, the selective distribution of lesions at bifurcations in the upper bronchial tree could result from both selective deposition and slowed clearance and in turn result in prolonged retention of high local concentrations of inhaled carcinogens. Although there is no direct evidence that ineffectual clearance contributes to the development of bronchogenic carcinoma, a causal relation has been suggested between adenocarcinoma, sites of local particle retention, and inadequate clearance in the nasal passages of furniture workers (Hadfield and Macbeth, 1971; Morgan et al., 1973).

Evidence is accumulating that dysfunction of bronchial clearance plays a role in the pathogenesis of chronic bronchitis; mucus transport is impaired in people who have the disease (Wanner, 1977). Cigarette-smokers and persons with chronic obstructive pulmonary disease show a wider variation in clearance rates than do non-smoking healthy people (Albert et al., 1973; Gongora et al., 1981). In the former groups, the within-subject variation is also great. That suggests that loss of control of mucociliary transport could either cause or result from chronic obstructive lung disease. It has also been shown in experimental animals that modest changes in mucociliary clearance rates are associated with secretory epithelial changes in small bronchi and bronchioles—changes that, if continued, could lead to clinical manifestations of bronchitis (Schlesinger et al., 1983; Gearhart and Schlesinger, 1987). Other studies with humans have suggested that mucociliary dysfunction is an early indication of pathogenic changes in the lungs. For example, retarded mucociliary clearance has been demonstrated in bronchitic people who showed no sign of airway obstruction (Mossberg and Camner, 1980), whereas young smokers with various degrees of impairment of tracheal mucus transport had no overt bronchitic symptoms and had normal results for pulmonary function tests of airway obstruction (Goodman et al., 1978).

The pathogenetic implications of alterations in clearance from the alveolar region have not been examined to the same extent as changes in mucociliary clearance. Alveolar clearance rates appear to be reduced in people with chronic obstructive lung disease and in cigarette-smokers (Cohen et al., 1979; Bohning et al., 1982); that suggests some relation between altered defense and disease development. Clearance dysfunction has also been shown in animals that have viral infection (Cresia et al., 1973).

The adequate performance of alveolar macrophages is critical to the effectiveness of lung defense in minimizing the residence time of deposited toxicants. For example, phagocytosis plays an important role in the prevention of particle entry into fixed tissues of the lung, from which clearance is very slow; accumulations of several types of dust have been directly linked to development of lung disease. Damage to macrophages has been implicated in the pathogenesis of chronic lung diseases involving proteolysis—e.g., emphysema, fibrogenesis, silicosis, and asbestosis (Brain, 1980; Warheit et al., 1984)—as well as in an increased risk of viral and bacterial infections (Hocking and Golde, 1979).

In vivo clearance studies are generally performed by examining the rate of removal or transport of tracer particles from the lungs as a whole or from specific individual airways. Serial sacrifice or fecal analysis techniques can be used with experimental animals. Measurements of rates or times are, however, strongly influenced by specific methods.

Local Mucus Velocity

Mucus transport velocities in the nasal passages, trachea, and main bronchi can be measured directly. The techniques involve monitoring of tracers on the epithelium, measurement of the movement of boluses of particles selectively deposited in these airways, or moving through them from more distal areas. Rates of movement are measured by determining the time needed for a bolus or tracer to traverse a calibrated distance or to move between two anatomically defined areas. The tracers can be cellular debris or material specifically introduced into the airways, e.g., India ink droplets, Teflon disks, radiolabeled resin beads, powders, colored solutions or dyes, or pollen grains. They are generally introduced into the trachea by an airstream or by instillation (both via bronchoscopy) or into the nasal passages by an airstream or simply by placement at the desired site.

Various measurement techniques have been used. For example, saccharin particles have been placed in the nose, and the time until the subject reported the first taste of sweetness measured (Proctor et al., 1977). Some nasal markers have been viewed directly in the nasopharynx after placement on the anterior nasal mucosa (Van Ree and Van Dishoeck, 1962; Bang et al., 1967). The most objective and sensitive procedures involve placement or inhalation of radioactive or radiopaque tracers into the upper respiratory tract or central airways and then viewing with external monitoring methods, including fluoroscopy (Friedman et al., 1977; Goodman et al., 1978; Mezey et al., 1978), cinebronchofiberoscopy (Sackner et al., 1973; Santa Cruz et al., 1974; Toomes et al., 1981), scintillation detection (Proctor et al., 1977; Man et al., 1980; R. K. Wolff et al., 1982), and gamma-camera imaging

(Quinlan et al., 1969; Chopra et al., 1977; R. K. Wolff et al., 1982). A modification involves measurement of transport velocities in the trachea or main bronchi by monitoring of the movement of boluses of radiolabeled particles, either inhaled in a manner that maximizes deposition in the central airways or coming from more distal bronchi. Movement along the trachea can be monitored with a gamma camera or scintillation detectors (Yeates et al., 1975; Foster et al., 1978, 1982; Schlesinger et al., 1978; R. K. Wolff et al., 1982).

The advantage of local-velocity techniques is that they allow measurement in anatomically defined airways. In addition, because a specific site is used, there is no question as to whether altered clearance rates due to toxicant exposure resulted from alterations in the mucus system or from a change in tracer-particle deposition pattern; the latter is a possibility when whole-lung clearance assays are used. However, there are a number of disadvantages. Many of the techniques are invasive, in that tracer particles are introduced into the airway of interest, the use of anesthetics might affect transport rates, and the procedure of introducing particles might result in trauma to the airways.

Alterations in nasal or tracheal transport rates have been used as markers of disease or of response to inhaled pollutants, because they are easier to measure than is whole-lung clearance (Sackner et al., 1978; Wolff et al., 1981; Majima et al., 1983; Stanley et al., 1985). However, findings in the upper respiratory tract or trachea cannot be extrapolated to the lower lung and might not be adequate indexes of overall respiratory tract effect. Although nasal and tracheal mucus clearance could be affected in any impairment of overall clearance, altered bronchial clearance due to pollutant exposure is often associated with no change in nasal or tracheal transport rates (Albert et al., 1974; Schlesinger et al., 1978, 1979; Leikauf et al., 1981). Thus, the use of alterations in clearance in specific airways as a marker of mucociliary function is valid only if the agent of interest or a disease affects the region being meas-

ured. In any case, neither nasal nor tracheal transport tests can provide direct assessment of mucociliary function in the small bronchi and bronchioles; however, it is important to note that mucociliary dysfunction is probably of greater consequence for pathogenesis than many other changes. To avoid problems in the respiratory system associated with regional measurements, whole-lung clearance can be used as a marker.

Whole-Lung Clearance

The technique most commonly used to measure whole-lung clearance involves inhalation of a radiolabeled tracer aerosol. Tracer materials used include Teflon, polystyrene latex, hematite, magnetite, and clay. The total amount of radioactivity remaining in the lungs at selected times is measured with external detector systems. The decline in rate of emission of radioactivity, corrected for radioactive decay, represents clearance. Unlike the aforementioned methods, this technique does not measure the movement of individual particles or boluses, so actual rates of transport are not obtained.

Various types and configurations of scintillation detectors have been used to measure lung clearance; they may be divided into scanning and stationary detector systems. In the former arrangement, either the detectors move relative to the thorax of the subject, who has inhaled the tracer aerosol (LaBelle et al., 1964; Holma, 1967a,b; Camner and Philipson, 1971, 1978; Camner et al., 1971), or the subject is moved relative to stationary detectors (Albert et al., 1968). Measurements are obtained when the subject is in preselected positions or during continuous scanning, i.e., during movement at a constant rate. Scanning systems are relatively independent of the apex-to-base distribution of deposited tracer particles in the lungs, in that they provide a longitudinal activity distribution map. They also reduce counting variations due to slight differences in positioning of the subject or due to subject movements, in that they depend less on the exact equivalent geometry of the subject's location in relation to the detectors than do stationary systems.

In stationary or fixed systems, collimated detectors are placed in positions relative to the subject's thorax. Configurations used include anterior placement of a single, central detector (Albert and Arnett, 1955; Toigo et al., 1963; Thomson and Short, 1969); placement of two detectors, one centrally and one laterally (Luchsinger et al., 1968); anteroposterior or bilateral placement of twin, axially opposed detectors (Booker et al., 1967; Thomson and Paria, 1974; Leikauf et al., 1981; Schlesinger et al., 1982, 1986); and use of multiple detectors (Albert et al., 1969; Stahlhofen et al., 1981; Bailey et al., 1982). Two other systems have also been used: the gamma camera (Lourenco et al., 1971b; Sanchis et al., 1972; Puchelle et al., 1982) and the whole-body counter (Cresia et al., 1973; Bohning et al., 1982; Snipes et al., 1983).

Unilateral detector systems require exact positioning for reproducible results, whereas dual or multidetector systems (in which the signal output is combined) are less sensitive to changes in position of the subject in the measurement plane or to effects of redistribution of particles in the lungs. The gamma camera permits assessment of total clearance and allows visualization of the distribution of retained particles at various times after exposure.

The airways within which the test aerosol is deposited are in a three-dimensional array in the lungs and are therefore at various depths relative to the detectors. Thus, the efficiency with which retained activity in each airway is measured varies. It follows that detector configuration affects the shape of the clearance curve. In addition, because of sensitivity differences, the amount of activity needed for successful analysis varies with the type of detector used. The gamma camera offers the greatest advantage in spatial resolution, but has low sensitivity and requires large amounts of radioactivity. Scanning systems have poorer spatial resolution, but better sensitivity. Multiple stationary detectors offer little information on intrathoracic particle distribution, but are the most sensitive. The most responsive fixed system is the whole-body counter; however, it is not suitable for use during the first few days after tracer exposure, because, not being collimated, it cannot effectively distinguish between activity in the lungs and that cleared into the stomach during the initial, rapid tracheobronchial clearance phase. But it can be used to monitor long-term clearance, once the activity in the rest of the body is lower than that in the lungs. Thus, some other technique must be used to monitor the mucociliary clearance phase and allow determination of the time when the lung is the only organ with appreciable remaining activity.

One of the major problems associated with external monitoring techniques is the dependence of the observed mucociliary clearance pattern on the pattern of initial deposition of the tracer aerosol. That dependence exists because the techniques are indirect and clearance characteristics are influenced by mucociliary transit rates. For example, an apparent increase in clearance rate after pollutant exposure could be due to a proximal shift in deposition of the tracer aerosol, rather than to an effect on the clearance system itself. That could be a special problem in the comparison of different groups; e.g., subjects with chronic obstructive lung disease tend to have greater central airway deposition of a given tracer aerosol than healthy subjects (Lippmann et al., 1980). Such differences must be borne in mind in the assessment of clearance changes as markers of exposure or disease.

The shapes of mucociliary clearance curves depend heavily on tracer-particle deposition. In studies of respiratory-region clearance, however, different clearance rates can also be associated with particles of different sizes (Bailey et al., 1982), because there can be size-dependent differences in macrophage phagocytosis, in vivo solubility, etc. Differences in long-term clearance of particles of equivalent size but consisting of different materials have been noted in both humans and experimental animals (Stahlhofen et al., 1981; Schlesinger et al., 1982).

The experimental assessment of clearance from the respiratory region requires that measurements be performed over, perhaps, several months. If radioactively tagged tracer aerosols are used, a nuclide having a relatively long half-life is required. In addition, because the total dose to a subject should be minimized, long counting times might be required for data to be statistically reliable. Thus, very long-term clearance studies that use humans could preclude use of radioisotopic tracers, because of potential health risks.

A technique that avoids the difficulties associated with the assessment of clearance is magnetopneumography (MPG). As discussed in Chapter 3, MPG allows direct assessment of lung burdens of magnetic materials in suitably exposed populations. In addition, humans and experimental animals have been exposed to inert magnetic dusts to assess clearance (Valberg and Brain, 1979; Cohen et al., 1979; Freedman and Robinson, 1981; Halpern et al., 1981). Measurement of the remanent field over time provides an index of clearance.

MPG techniques have some advantages over the radioaerosol techniques, with respect to both temporal resolution and spatial resolution in the measurement plane. In addition, magnetic techniques have potential advantages, in that they yield some information that cannot be obtained with other clearance techniques. The time dependence of decay of magnetic field after the external field is applied is affected by the viscosity of the medium in which the deposited particles reside (Williamson and Kaufman, 1981). Therefore, as free particles are translocated from alveolar surfaces or engulfed by macrophages, the response to an externally applied field can change. When magnetic measurements are used to assess clearance over a long period, some or all of any observed decline in remanent moment could be due to immobilization of particles. In addition, the hysteresis curves of magnetization and relaxation (loss of magnetic alignment) can provide information on the amount of fibrosis in lung tissue (Cohen, 1975).

One technique used for whole-lung clearance in both humans and experimental animals allows visualization of tracer particle distribution without the need for radiolabeled aerosols. It involves insufflation of radiopaque tantalum powder through a tracheal catheter. Serial chest x-ray pictures are then taken over a period of months, to provide visualization of clearance, which is measured on the basis of visual scoring of film intensity (Gamsu et al., 1973; Wood et al., 1973). The technique provides some measure of whole-lung and regional clearance patterns and allows measurement in anatomically defined airways, but it is invasive, is only semiquantitative, and requires high radiation doses, so it is unsuitable for routine use in humans. It also requires several grams of tantalum, which might overload clearance systems. The technique therefore has not found widespread use.

INJURY TO AIR-BLOOD BARRIER

Conducting-Airway Permeability

The epithelium of the conducting airways is normally fairly impermeable, so material deposited on airway surfaces is absorbed slowly. This protective barrier depends on the integrity of the tight junctions between epithelial cells. Exposure to ozone (Kehrl et al., 1987) and other atmospheric agents can alter barrier function, increasing penetration of inhaled materials into the blood. Hyperpermeability has been demonstrated in the absence of morphologic alteration after exposure of experimental animals to a number of noxious agents, including cigarette smoke, nitrogen dioxide, and zinc oxide (Simani et al., 1974; Boucher et al., 1980; Ranga et al., 1980; Hulbert et al., 1981). Thus, assays of epithelial permeability can be used to screen for, and act as markers of, epithelial damage. However, the damage assessed is nonspecific.

Hyperpermeability has pathophysiologic implications (Boucher, 1980). A decrease in the ability of the epithelium to act as a barrier might result in an increase in the translocation of inhaled, deposited materials through the airway wall. Increased loading of antigens, for example, could increase immunologic responsivity. Hyperpermeability could also result in an increased flow of bronchoactive agents to effective sites, and thus produce hyperreactivity (Boucher et al., 1977a,b, 1979). Hyperpermeability might alter the efficiency of mucociliary transport by changing ionic or macromolecular transport across the epithelium; permeability changes could alter mucus hydration and viscosity. Epithelial permeability might be assessed by examining ion transport or by measuring the passage of specific molecules across the epithelium.

Epithelial Ion Transport

Airway epithelial cells exhibit active ion transport. In concert with submucosal gland secretions, this capability contributes to regulation of the volume and composition of the liquid lining of airway surfaces. Fluid absorption, driven by active sodium transport, is the dominate basal solute flow across the surface of proximal human airway epithelia (Knowles et al., 1984). It eliminates the necessity for a large reduction in the volume of airway surface liquid as it is moved by ciliary activity from distal sites of large aggregate surface area to proximal sites of smaller surface area. Whereas fluid is absorbed under basal conditions, fluid secretion driven by active chloride transport can be induced in the presence of a favorable electrochemical gradient and an apical cell membrane that is permeable to chloride.

The flux of ions across the airway epithelium generates a transepithelial potential difference (PD), which reflects the magnitude and direction of active ion transport and the passive ion permeabilities of cellular and paracellular pathways; PD differs by site in the respiratory tract (Knowles et al., 1981). Some sites might be better than others for assessing effects of pollutant exposure. The nasal epithelium is a particularly attractive model for studying toxic effects, because of its accessibility; the ability to perform parallel in vitro studies in freshly excised or cultured epithelial cells; the availability of techniques, such as the use of nasal filter paper or nasal wash, to obtain cell markers, mediators, or a measurement of the concentration of injurious agents; and the ability to obtain nasal epithelial cells for more direct assessment of changes in cell structure or biochemical function.

The measurement of epithelial PD in vivo has been used to explore normal epithelial function and dysfunction in acquired and genetic disorders (Knowles et al., 1983). In addition, some information regarding use of in vivo PD as a marker of disease due to pollutant exposure has been obtained. For example, the tracheal PD of a group of young asymptomatic smokers was reduced (Knowles et al., 1982); thus, the tracheal PD could indicate epithelial damage before symptoms are noted. Some of the individual values lie within the range of normal biologic variability. Nasal PD can be reduced in the presence of inflammatory processes.

A number of other agents have been shown to alter airway epithelial permeability and inhibit PD in vivo (Boucher, 1981; Stutts and Bromberg, 1987). They include ozone, nitrogen dioxide, and sulfur dioxide. Some agents tested in vitro—e.g., zinc, mercury, and formaldehyde—have induced similar responses (Stutts et al., 1981, 1982, 1986). Therefore, changes in PD are unlikely to serve as markers of specific exposures, or to provide accurate information on exposure concentration.

Although measurement of PD might serve as an index of exposure or response to a pollutant, the mechanism of the effect on epithelial barrier function cannot be ascertained from this measurement alone. It is difficult to differentiate between a change in ion current and a change in tissue resistance. If there is no change in PD, one cannot be sure that several barrier properties, such as ion transport and tissue resistance, were not altered in an equal and opposite fashion. The ability to estimate epithelial resistance in vivo would be useful; such measurements appear feasible and might improve the ability to characterize the nature of any insult (Knowles et al., 1986).

The ability to perform more carefully controlled experiments with fresh tissues in vitro might also allow assignment of an observed effect to a change in ion current or to a change in cellular or paracellular resistance. Potential limitations to parallel in vivo and in vitro studies are related to dosimetry, blood flow, and mucus barriers. The preservation of normal ion transport in primary cultures of mammalian (including human) pulmonary epithelia and the expression of genetic dysfunction in epithelial cells (as in cystic fibrosis) suggest that further assessment of toxic exposures could be pursued in parallel in epithelial cell culture systems and in vivo. Thus, ion transport characteristics of the airway epithelial barrier might represent useful biologic markers to monitor the effects of acute and chronic exposures to pollutants.

Molecular Tracer Procedures

A change in permeability can be assessed by examining the transepithelial transport of molecules into the blood. A good molecular tracer should have several properties:

- It should be nontoxic.
- Its movement across the epithelium should be the rate-limiting process in its clearance from the lungs. The marker should be cleared rapidly from the lungs by the pulmonary, bronchial, or lymphatic circulations, and its clearance should not be limited by its movement through the interstitial matrix. Otherwise, a change in blood flow due to edema or lung disease could affect its clearance independently of changes in epithelial permeability.
- Its route of clearance should be known. An accurate interpretation of the clearance data with respect to lung injury is difficult, unless it is known whether increased clearance is due to opening of the gaps between epithelial cells, an increase in vesicular transport, or simply an increase in the epithelial surface area for diffusion.
- Its clearance should increase only in the presence of lung injury. The marker must be able to distinguish lung injury from the normal changes in a noninjured lung, such as an increase in lung volume, an increase in blood flow, or an increase

in interstitial fluid volume due entirely to high lung vascular pressures.
- It must not be transformed in the lung. The marker must not be denatured by the delivery process or degraded in the lung airspaces or interstitium. In addition, if a radioactive tag is used, the binding between the isotope and the marker must be tight enough to prevent its separation.
- Its kinetics must facilitate measurement. The marker must clear rapidly enough to allow accurate measurement of its clearance within approximately 1 hour for clinical studies, but it must not clear so rapidly that the time required for deposition interferes with analysis.
- It should be easily detected. A marker that requires numerous blood samples or labor-intensive assays would make measurement too cumbersome, invasive, or costly.

Those criteria apply to tracers used for assessing conducting-airway permeability, as well as alveolar epithelial permeability.

Two basic techniques have been used to assess conducting-airway permeability with macromolecular markers. The first involves use of horseradish peroxidase (HRP), a glycoprotein with a molecular weight of approximately 40,000 daltons. HRP is instilled into the trachea, and epithelial permeability is assessed by measuring plasma concentration of HRP at various times after instillation with radioimmunoassay or enzyme-linked immunoassay; electron microscopy of tissues allows localization of HRP for assessing routes of movement (Conner et al., 1982; Simani et al., 1974; Hulbert et al., 1981; Ranga et al., 1980; Boucher et al., 1978, 1980). The procedure is used in experimental animals, and the effects of inspired pollutants can be assessed by comparing test with control animals; in normal animals, HRP transfer rates are very low (Hogg et al., 1979). The finding of HRP in intercellular spaces when concentrations in the blood are increased indicates that the blood assays actually measure a loss of barrier function. Studies with cigarette smoke in animals have suggested that the exudative phase of the inflammatory response is associated with hyperpermeability and that function returns to normal when the repair phase begins (Hulbert et al., 1981). In addition, changes in tracheal permeability are related in a dose-dependent fashion to dose of cigarettes; effects are seen in guinea pigs after as few as 20 puffs (Boucher et al., 1980). It must be borne in mind, however, that correct interpretation of plasma HRP requires knowledge of the HRP clearance rate from plasma; any toxic effect on this rate could affect interpretation of the permeability assay (Conner et al., 1985).

Another procedure to assess tracheobronchial airway permeability (and one that is used in humans) involves inhalation of radioactively tagged (^{99m}Tc) aerosols of diethylenetriaminepentaacetate (DTPA), which has a molecular weight of 492 daltons, and its analysis in the blood or with thoracic scanning to obtain a clearance half-time. With proper adjustment of particle size and breathing characteristics, the inhaled aerosol can be preferentially deposited on the tracheobronchial tree with minimal respiratory-region (alveolar) deposition (Oberdörster et al., 1986). Clinical studies have yielded a wide range of reported half-times of DTPA (Jones et al., 1980; Rinderknecht et al., 1980; Mason et al., 1983; Kennedy et al., 1984; O'Byrne et al., 1984); some of the variation could be due to differences in deposition sites of the aerosol in different studies and differences in absorption at these sites.

Cheerma et al. (1988) examined the diffusion and binding characteristics of DTPA in measuring permeability across alveolar epithelium and bronchial mucosa. They found that, because DTPA has a high affinity for the mucosal layer of the bronchial epithelium, it might not be suitable for measuring tracheal and bronchial clearance, but is more useful for measuring alveolar clearance.

Thus, there is a need to standardize the assay for particle size and breathing characteristics, if it is to be a reliable marker of response. In addition, the sensitivity of the procedure depends on the site of action of the pollutant being assessed. For example, rats exposed to ozone showed a greater and more persistent response in the bronchoalveolar zone than in the trachea, because the site of major ozone deposition was the bronchoalveolar zone (Bhalla et al., 1986).

Alveolar Epithelial Barrier

The layer of epithelial cells lining the airspaces of the lungs forms a tight barrier that greatly restricts the movement of most solutes. It is not clear what happens when lung injury causes an increase in the permeability of the barrier, but two responses are possible: enzymes gain access to the lung tissue and cause increased tissue damage; and solutes and fluid pass more easily from the interstitial spaces and vascular spaces resulting in either an alteration in the composition of the liquid lining of the airspaces or alveolar edema. The specific response to increased permeability undoubtedly depends on the location and on the size of the solute to which the epithelium has become more permeable.

The study of the clearance of tracers from the alveoli is complicated by the diverse and numerous obstacles between the airspaces and the blood. Tracers can move by diffusion or vesicular transport from the alveoli, across the epithelium and endothelium, and into the blood (Path 1, Figure 3-2). However, in regions where the basement membranes of the epithelium and endothelium are not closely apposed or in conditions where blood vessels are not perfused, a tracer can diffuse or flow through the interstitium into a perfused blood vessel or a lymphatic vessel (Paths 2a and 2b, Figure 3-2) (Peterson and Gray, 1987). It can also diffuse or flow into a fluid reservoir in the interstitium, such as those which develop around airways in the presence of interstitial edema (Path 2c, Figure 3-2) (Havill and Gee, 1984). The formation of those reservoirs can complicate the measurement of epithelial permeability, because they can prevent the tracer from leaving the interstitial spaces. Macklin (1955) proposed the existence of a mechanism for clearance of particulate matter and large solutes via "sumps" found at bronchioles;

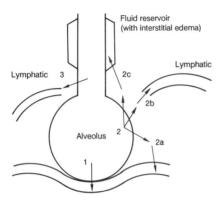

FIGURE 3-2 Clearance pathways for markers in the airspaces.

the sumps allow the solutes to flow by convection from the airspaces directly into the lymphatic vessels (Path 3, Figure 3-2). The existence of a variety of pathways and mechanisms for the movement of a tracer from the airspaces suggests that clearance of a tracer can be complicated by simple changes in the activity of a normal transport mechanism.

The most commonly used marker of changes in alveolar epithelial permeability is DTPA. Its clearance rate has been found to increase in response to a variety of conditions and insults in humans, including interstitial lung disease (Rinderknecht et al., 1980), ozone exposure (Kehrl et al., 1987), sarcoidosis (Dusser et al., 1986a), and cigarette-smoke exposure (Dusser et al., 1986b). Animal studies have also shown that DTPA clearance increases in response to inhaled cigarette smoke (Minty and Royston, 1985), ozone (Bhalla and Crocker, 1986), and irradiation (Ahmed et al., 1986).

Some problems are evident in the use of ^{99m}Tc-DTPA as a marker of alveolar epithelial injury. Its clearance rate increases with lung inflation (Rinderknecht et al., 1980; Peterson et al., 1986; Rizk et al., 1984). The ^{99m}Tc label can dissociate from DTPA in the presence of oxidants (Nolop et al., 1986); this could be a problem in assessing response to some pollutants. Because free technetium clears more rapidly than does ^{99m}Tc-DTPA, dissociation of the label would yield an increased clearance rate (Egan, 1980); that could explain

the increased clearance rate measured in cigarette-smokers (Nolop et al., 1986). The use of [113]In-DTPA has been proposed, because indium is more tightly chelated by DTPA than is technetium (Nolop et al., 1986). Finally, although the relatively rapid clearance of DTPA from the lungs of healthy subjects and those with lung injury allows accurate measurements of clearance within 30 minutes, it might also cause a measurable background concentration to appear, because of recirculation of the tracer.

Other molecular tracers have been investigated. Clearance of instilled tracers with an Einstein-Stokes radius greater than 2 nm from the airspaces might not be affected by lung inflation (Egan, 1980), so the use of a tracer larger than DTPA (radius, 0.6 nm) might overcome one potential problem with DTPA. Bovine serum albumin (radius, 3.6 nm) has been labeled with [99m]Tc (Hnatowich et al., 1982; Peterson et al., 1989). In anesthetized sheep, the albumin clearance rate measured with nuclear imaging increases in the presence of lung injury, but is unaffected by changes in lung volume or by lung edema due to increased lung vascular pressures (Figures 3-3, 3-4, 3-5). However, the process of labeling and aerosolizing the albumin could cause the formation of macroaggregates, which complicate interpretation of the data. Furthermore, albumin might leave the airspaces by specialized transport mechanisms (K. J. Kim et al., 1985). Mannitol has also been proposed as a marker of airway clearance (Taylor et al., 1983), but its use might suffer from some of the problems associated with using DTPA, in that mannitol is small (radius, 0.3 nm).

In summary, the search for an accurate and useful technique for the in vivo measurement of changes in epithelial permeability is worthwhile, but has not been totally successful. It might be necessary to use simpler experimental models of lung epithelium—e.g., isolated perfused lungs or monolayers of cultured epithelial cells—to identify markers before that can be used in vivo.

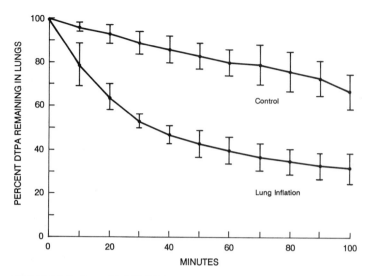

FIGURE 3-3 10 cm H_2O PEEP increases DTPA clearance rate. Šource: Reprinted with permission from Peterson et al., 1988.

Vascular Injury

Endothelial cells found in pulmonary blood vessels function to retard passage of fluid, protein, and some other blood components from the vessel lumen into the interstitium and the airspaces of the lung. In addition to that barrier function, pulmonary vascular endothelium performs such functions as the removal or metabolism of endogenous and exogenous circulating agents and the synthesis of biologically active substances (e.g., prostacyclin and factor VIII antigen) that help to maintain vascular homeostasis. In health and disease, the endothelium interacts with blood cells (such as platelets and leukocytes) and with cells that form the vasculature (such as fibroblasts, pericytes, and other cells of the interstitium), as well as with smooth muscle cells in precapillary portions of the vascular tree.

Those qualities change when endothelial cells are injured. In theory, all those and perhaps other characteristics of endothelial cells can be exploited as markers of injury. The purpose of this section is to review and to comment briefly on currently used indicators of endothelial injury and to provide an example of a specific functional characteristic of endothelium that might constitute a useful biologic marker of injury.

Chemically Induced Injury to Endothelium

Structural alterations in pulmonary endothelium have been demonstrated experimentally after toxic insult with a number of chemicals, including such diverse agents as α-naphthylthiourea, papain, ethchlorvynol, bleomycin, bromocarbamide, iprindole, some gases (e.g., oxygen), epinephrine, and the pyrolizidine alkaloid plant toxins (Witschi and Côté, 1977). Blebbing and swelling of endothelium occur with each of those toxicants. The changes in endothelium are usually associated with changes in other lung cell types. However, the exact temporal relationship between injury to endothelium and changes in other cell types varies from one toxicant to another. For example, after administration of bleomycin, iprindole, or oxygen, changes in endothelium are seen before changes in other cell types; after intratracheal administration of papain, changes in alveolar Type II cells and fibroblasts are seen before changes in endothelium; after administration of other toxicants,

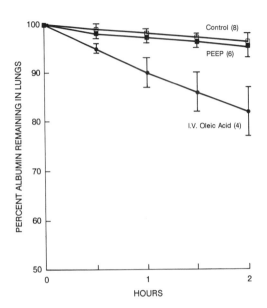

FIGURE 3-4 Albumin clearance discriminates between lung injury and lung inflation.

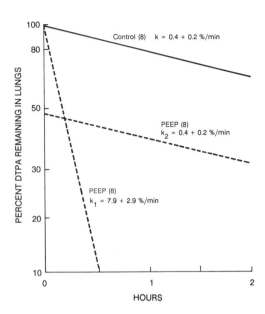

FIGURE 3-5 Compartment analysis of DTPA clearance.

such as epinephrine, changes in endothelium and changes in other cell types seem to occur simultaneously.

The location of the earliest damage to pulmonary endothelium might also vary from one toxicant to another. For example, in oxygen toxicity, capillary endothelium is the first to show changes; after bleomycin administration, the earliest damage is to arterial and venous endothelium. Thus, markers that could distinguish injury in various parts of the vascular bed might provide important clues to mechanisms and risks associated with various pulmonary diseases.

Loss of Endothelial Barrier Function

A major role of endothelium is to prevent loss of fluid from vessel lumina. Endothelial cell injury is usually manifested clinically as evidence of pulmonary edema. Pulmonary edema is classified into two types. In hydrostatic or hemodynamic edema, abnormally high intravascular pressures in small parenchymal vessels lead to flux of fluid from them. In permea-

bility edema, intravascular pressures can be normal, but leaks in alveolar capillaries allow increased flux of water and protein into the extravascular compartment. The former type does not always entail injury to the vascular endothelial barrier, but can arise, for example, as a result of constriction of pulmonary venules. The most common examples of hemodynamic pulmonary edema result from chronic left-sided heart failure or mitral valve disease (Fishman, 1980). Permeability edema is associated with formation of a protein-rich lymph that arises from an injured endothelial barrier that allows increased passage of plasma proteins into the interstitium of the pulmonary parenchyma. Permeability edema results from exposure to some noxious airborne agents, such as nitrogen dioxide, and also occurs in acute respiratory diseases. In animal studies, it also results from a number of chemical insults to pulmonary capillary endothelium, e.g., after exposure of the pulmonary vasculature to oleic acid, alloxan, α-naphthylthiourea, or phorbol ester.

In animals, permeability edema has been measured on the basis of protein leakage from the pulmonary vasculature. Lung lymph flow, lymph fluid protein concentration, and accumulation in excised lungs of radio-labeled protein introduced into the blood all can be measured. They have provided much important information on vascular leak in laboratory animal studies, but not directly in humans.

Loss of barrier functions is also reflected in an increase in extravascular lung water. Regardless of the cause, excess extravascular water can be detected with a variety of methods. In general terms, those methods can be divided into invasive and destructive methods, invasive and nondestructive methods, and noninvasive and nondestructive methods (Table 3-2). The destructive techniques have the major advantage of accuracy and thus are often reported as the putative "gold standard" by which other techniques are judged (Staub, 1974; 1986). Clearly, however, they are unsuitable for clinical studies.

The nondestructive techniques, although useful in a clinical setting, suffer to various degrees from inaccuracy, nonspecificity, impracticality, and expense. For instance, pulmonary edema can be diagnosed clinically in a patient with a characteristic history (e.g., acute onset of dyspnea and tachypnea) and physical findings of cyanosis and rales during chest auscultation (Staub, 1974, 1986). However, other conditions can cause similar findings. In any case, the approach is nonquantitative and relatively insensitive to smaller accumulations of extravascular water. Nonetheless, because of its simplicity and lack of expense, the clinical examination remains an important means for detecting the presence of acute pulmonary edema.

Of the other available techniques, four deserve special consideration: chest roentgenography, the indicator-dilution method, and the newer techniques of positron-emission tomography and nuclear magnetic resonance.

Chest roentgenography has many favorable features for use as a marker of lung injury. It is practical, widely available in a variety of useful settings, and relatively inexpensive. Its accuracy and sensitivity in detecting pulmonary edema are disputed. When strict attention is paid to technical factors, some have argued that the chest roentgenogram is quite sensitive to changes in lung water content, and accurate inferences can be made about the magnitude of such changes (Pistolesi and Guintini, 1978; Milne et al., 1985). However, several other groups have tested it against presumably more accurate tech-

TABLE 3-2 Methods for Detecting Excess Extravascular Lung Water Accumulation

Category	Methods
Invasive and destructive	Gravimetrics (Staub, 1974, 1986)
	Histology (Staub, 1974, 1986)
Invasive and nondestructive	Indicator dilution (Baudendistel et al., 1982; Grover et al., 1983; Sibbald et al., 1983; Eisenberg et al., 1987; Sivak and Wiedemann, 1986; Effros, 1985; Lewis et al., 1982)
Noninvasive and nondestructive	Clinical examination (Staub, 1974, 1986)
	Pulmonary mechanics (Staub, 1974, 1986)
	Chest roentgenography (Milne et al., 1985; Pistolesi and Guintini, 1978; Baudendistel et al., 1982; Grover et al., 1983; Sibbald et al., 1983; Eisenberg et al., 1987; Sivak and Wiedemann, 1986)
	Soluble-gas uptake (Overland et al., 1981)
	Microwave transmission (Iskander et al., 1979)
	Compton scatter (Loo et al., 1986)
	X-ray computed tomography (Hedlund et al., 1984; 1985)
	Positron-emission tomography (Schuster et al., 1985; Rhodes et al., 1981; Wollmer et al., 1984; Schober et al., 1983; Schuster et al., 1986; Cutillo et al., 1984)
	Nuclear magnetic resonance (Cutillo et al., 1984; Morris et al., 1985; Wexter et al., 1985)

niques and have not been able to demonstrate an acceptable degree of accuracy (Baudendistel et al., 1982; Grover et al., 1983; Sibbald et al., 1983; Sivak and Wiedemann, 1986; Eisenberg et al., 1987).

Two techniques, indicator dilution and positron-emission tomography, measure the intravascular components of lung water. Thus, they can measure extravascular lung water accumulation, which is in fact the entity of interest, inasmuch as the abnormal accumulation of extravascular water represents breakdown in endothelial cell barrier function. In addition, gas rebreathing techniques have been used to estimate lung tissue volume and pulmonary capillary blood volume. From those two volumes and estimates of ratios of wet to dry weight of tissue and blood, intravascular and extravascular water can be estimated.

Indicator-dilution methods of measuring extravascular lung water are based on the concept that the mean transit time of an indicator through a fluid depends on indicator flow rate and the volume of the fluid (Lewis et al., 1982; Hedlund et al., 1984; Sivak and Wiedemann, 1986). For a given flow rate, if volume is small, the mean transit time will be small, and vice versa. To measure extravascular lung water, two indicators are used: one that can diffuse through the entire lung water volume and one that is limited to the intravascular, nondiffusible volume. Although a number of indicators have been used, the two that have achieved the greatest acceptance are heat (actually, temperature change) as the diffusible indicator and dye (e.g., indocyanine green) as the nondiffusible indicator. The green dye binds immediately in vivo to albumin and thus remains intravascular during the period of lung water measurement. Extravascular lung water (EVLW) can be calculated as

$$EVLW = CO\,(MTT_t - MTT^{gd}),$$

where CO is the cardiac output (i.e., a measure of vascular flow) and MTT is the mean transit time of the thermal (t) or green dye (gd) indicator. This method,

called the thermal-green dye double-indicator dilution technique, has been verified by numerous groups as accurate (in most, although not all, instances in which EVLW is increased), reproducible, and reasonably sensitive to changes in EVLW (i.e, it will reliably detect approximately a 20% change) (Sivak and Wiedemann, 1986). Nonetheless, it is moderately invasive (catheters in the pulmonary and femoral arteries are required) and thus is not suitable for general population screening studies.

Positron-emission tomography (PET) is a nuclear-medicine technique that produces quantitative tomographic images of the tissue distribution of a previously administered positron-emitting radionuclide. It uses image reconstruction algorithms identical with those used during routine x-ray computed tomography (CT). However, unlike x-ray CT, which cannot distinguish between intravascular and extravascular water (Hedlund et al., 1984, 1985), PET measurement of EVLW is feasible because it subtracts the intravascular water content (IVW) of a region from the total lung water content (TLW) of the same region (Schuster et al., 1985). Unlike the indicator-dilution method previously described, PET is less sensitive to errors caused by the underestimation of EVLW in poorly perfused areas of lung.

The intravascular component of EVLW is measured during PET by scanning the subject at least 2 min after inhalation of ^{15}O-labeled carbon monoxide, a gas that avidly binds to hemoglobin. IVW is calculated by comparing the radioactivity in a given lung region with activity in blood samples taken during the scan. A similar procedure is used to measure TLW, except that the scan is obtained during equilibrium of the bolus infusion of ^{15}O-labeled water. The calculation of extravascular water content of a region involves the subtraction of IVW from TLW. Alternatively, a constant infusion of ^{15}O-labeled water, or ^{11}C- instead of ^{15}O-labeled carbon monoxide or density measurements instead of TLW measurements may be used (Rhodes et al., 1981; Schober et al., 1983; Wollmer et al.,

1984). Recent studies in whole animals have suggested that PET provides measurements of EVLW in both normal and edematous lungs with acceptable accuracy and is sensitive to small changes in EVLW after physiologic intervention (Schober et al., 1983; Schuster et al., 1986). Values obtained in humans have been comparable with those obtained in experimental animals.

PET appears to be ideal for measuring *regional* EVLW content. Because of technical problems associated with radioactivity counting in heterogenous tissues, whole-lung values for EVLW are more difficult to obtain. More important, however, are the cost and impracticality of PET as a clinical tool in that a scanner, a computer, a cyclotron, and several highly trained personnel are required for obtaining the measurements.

Proton nuclear magnetic resonance (NMR) imaging is a new, complex, and expensive technique for evaluating lung water content (Cutillo et al., 1984; Wexter et al., 1985). NMR depends on the electromagnetic properties of nuclei of some atoms that cause them to act like small, spinning bar magnets when placed in a strong magnetic field. The most abundant of those atoms is hydrogen, which contains one proton. The proton is the principal nucleus used in current magnetic resonance imaging experiments. When it is placed in a strong magnetic field, there is a slight net orientation of the protons along the magnetic field direction. The introduction of a radiofrequency (RF) excitation at a frequency specific for both the magnetic field strength and the protons under consideration causes the reorientation of the protons; when the RF excitation is removed, the protons return to their original orientation. That process (i.e., return, or relaxation, of the proton) emits RF energy, which is detected by a sensitive antenna or coil, amplified, and processed by a computer. The computer processing of space- and time-dependent RF emission creates an image of the concentration (i.e., density) and environment of protons in fat and water of soft tissue. Desirable features of proton NMR imaging are that no ionizing radiation is necessary and there are no bone artifacts in the image.

Although NMR imaging will probably yield the most accurate in vivo measurement of lung-water distribution, subtraction of the vascular component remains difficult. That problem, signal-to-noise ratio characteristics in the imaging of lung tissue, and the complexity of the technology as a whole make NMR imaging, like PET imaging, unlikely candidates for screening general populations for evidence of endothelial lung injury.

It is probably unwise to use lung-water measurements obtained with any technique to evaluate early lung injury. The abnormal accumulation of excess lung water represents not only a failure of endothelial barrier function, but also a failure of various other mechanisms (the most important of which is lymphatic function) that the lung can use to maintain normal water homeostasis. More useful as a marker of early injury would be a technique that detected breakdown of endothelial barrier function itself. Several groups have measured the flux of radiolabeled proteins across the pulmonary endothelium with external radiation detectors of various sorts (Gorin et al., 1978; Mintun et al., 1987). Although those techniques indeed seem to be more sensitive markers of lung injury than is the measurement of EVLW, they are still too new for prediction of how accurate, reproducible, and practical they will be in detecting lung injury in groups of humans.

In summary, no ideal means exists, or is likely to exist in the near future, for the detection of lung endothelial injury on the basis of either lung-water or capillary protein-flux measurements. The techniques that are simple, inexpensive, and practical to apply to large groups of humans are generally nonspecific and insensitive. The techniques that improve on specificity and sensitivity suffer in being expensive, impractical, and complex. The choice of method will depend largely on the specific goals of the program involved.

Nonbarrier Properties of Endothelium

As noted above, the nonspecificity, insensitivity, invasiveness, requirements for sophisticated equipment, and expense of currently available measures of endothelial barrier function limit their usefulness in diagnosing early permeability defects or subtle endothelial cell injury that can be associated in some people with a predisposition to serious pulmonary vascular disease. Obviously, markers associated with subtle, early defects in pulmonary endothelium that are sensitive, specific, and minimally invasive could be useful in identifying people at risk. Similarly, predisposition of people to diseases associated with defects in the endothelium might be predicted and such diseases prevented more effectively. For example, diverse types of trauma result in adult respiratory distress syndrome (ARDS) in some patients. Markers to identify subtle changes in endothelium might aid in identifying patients at risk of developing ARDS and in understanding its pathogenesis.

Research during the last several years has led to the identification of several non-barrier functions of pulmonary endothelium. From the standpoint of increasing our knowledge of mechanisms of lung injury, there is a need to understand better both barrier and nonbarrier functions of endothelium, to attain the capacity to assess them, and to determine how nonbarrier functions of endothelium are correlated with barrier properties. It should be recognized that changes in nonbarrier functions of endothelial cells might be useful predictors of deficits in the barrier function of the endothelium.

Metabolic Activity of Endothelium

The pulmonary vasculature performs a number of potentially important nonbarrier functions, some of which involve the modification of circulating concentrations of naturally occurring, biologically active substances, as well as drugs. Because the lung has a large vascular surface area and receives all of the cardiac output, it is uniquely situated to alter rapidly the circulating concentrations of vasoactive agents before they reach the arterial circulation. The capacity of the lung to clear the circulation of chemical agents and the potential importance of this function have been the subject of several reviews (Gillis and Pitt, 1982; Roth, 1985).

The ability to remove and metabolize substances reflects properties of endothelial cells of small vessels and capillaries in lung. For example, carrier-mediated transport of biogenic amines, such as 5-hydroxytryptamine (5HT) and norepinephrine (NE), into pulmonary vascular endothelium occurs. Available evidence indicates that 5HT and NE are taken up at different sites at the endothelial surface. After removal by the lung vasculature, those amines are metabolized by enzymes like monoamine oxidase and catechol-O-methyltransferase. However, the rate-limiting step in their initial removal from the circulation is transport from the vascular space, rather than intrapulmonary metabolism.

Circulating adenine nucleotides (adenosine monophosphate, adenosine diphosphate, adenosine triphosphate) are also altered on passage through the lung. Adenosine triphosphate, for example, does not survive passage through the pulmonary circulation. Biochemical and cytochemical studies have shown that, when those nucleotides are perfused through isolated lungs, all the radioactivity entering the pulmonary circulation is recovered in the effluent, but none remains in the form of the adenine nucleotide. The mean transit time and volume of distribution of those nucleotides are the same as those of intravascular markers. This indicates that the adenine nucleotides are metabolized in the pulmonary circulation without leaving the vascular space. Cytochemical data confirm that, although several cell types and organelles have phosphate esterases that hydrolyze nucleotides, only the enzymes that face the vascular lumen are exposed to and metabolize them. The location of the enzymes along the vascular lumen accounts for the fact that the metabolic products of adenine nucleotides appear in the venous circulation with no delay or tissue uptake.

The lung is also capable of hydrolyzing circulating peptide hormones, such as bradykinin and angiotensin I. Bradykinin is nearly quantitatively converted to shorter peptides in a single pass through the pulmonary circulation. The peptide is not taken up by lung, and its mean transit time and volume of distribution in perfused lung preparations are identical with those of intravascular markers, such as indocyanine green or blue dextran. Similarly, angiotensin I is extensively converted to angiotensin II on passage through the pulmonary circulation. Angiotensin-converting enzyme is located on the luminal surface of pulmonary endothelium; indeed, immunohistochemical studies have confirmed pulmonary endothelium as the only site of angiotensin-converting enzyme in the lung.

Studies in animals have revealed that impaired pulmonary metabolic function results from exposure to numerous toxicants. However, structural injury to pulmonary endothelium is not always associated with deficits in each type of metabolic function. For example, the pyrolizidine alkaloid, monocrotaline, produces pulmonary endothelial injury experimentally that is associated with reduced intrapulmonary clearance of 5HT by isolated lungs from treated animals (Roth, 1985). However, 5'-nucleotidase and angiotensin-converting enzyme activities in isolated lung preparations are apparently unaffected by treatment of rats with monocrotaline. Thus, chemically induced damage to lung might affect some functions of endothelium without altering others. This suggests some specificity in the endothelium-damaging action of some toxicants.

A number of studies have suggested that pulmonary metabolic functions may provide sensitive markers of endothelial injury. For example, exposure to the herbicide paraquat results in pulmonary lesions in humans and experimental animals. In rats, marked structural changes in alveolar epithelium have been commonly observed after paraquat administration, but alterations in vascular endothelium are much more subtle and infrequent. A modest but reproducible decrease in the ability of isolated lungs from paraquat-treated animals to remove perfused 5HT has been reported (Roth, 1985). The demonstration of impairment in 5HT clearance resulting from a treatment that produces little, if any, structural alteration in endothelium suggests that the pulmonary metabolic function could be a sensitive index of damage to pulmonary endothelium under some circumstances. That view is supported by studies of oxygen toxicity. Structural alterations in pulmonary capillary endothelial cells are an early manifestation of exposure to oxygen at 1 atmosphere. Block and Fisher (1977) reported that, although ultrastructurally demonstrable endothelial damage is not apparent until 48 hours of exposure to 100% oxygen, exposure for as little as 18 hours produces a significant decrease in 5HT clearance by lungs of exposed animals.

Those studies of pulmonary metabolic function in animals were performed in isolated lung preparations. The functions have also been studied in vivo both in animals and in humans with the multiple-indicator-dilution techniques described previously. For example, angiotensin-converting enzyme activity in the pulmonary vasculature has been studied with the synthetic substrate ^{3}H-benzoyl-phe-ala-pro (BPAP). BPAP and an intravascular marker are injected intravenously as a bolus, and the concentrations in the arterial (i.e., postlung) blood are compared over time. With this technique, the fraction of BPAP metabolized in a single passage through the pulmonary vasculature can be calculated.

There are potential pitfalls in using that and related methods to assess pulmonary microvascular injury (Stalcup et al., 1982). For example, pulmonary metabolic function can be influenced by changes in transit time and by inhomogeneity of perfusion, edema, and other factors that affect vascular surface area. When exogenously administered, radiolabeled substrates (e.g., ^{3}H-BPAP) are used, it is possible for endogenous substrates (e.g., angiotensin I and bradykinin) to compete with the tracer for metabolism and thereby confound interpretation of results. Furthermore, the lung might simultaneously synthesize and release the same test substance being removed or metabolized, and that would make interpretation of pulmonary extraction data difficult. In addition, questions have been raised about how to normalize metabolism data (e.g., whole lung vs. per unit lung weight, protein, DNA, etc.).

Careful monitoring of perfusion, intravascular pressures, and ventilation aid somewhat in ensuring reliability of data but do not resolve many of the potential problems. As mentioned above, the choice of a specific metabolic function and substrate can be of critical importance with regard to usefulness of a metabolic process in assessing lung microvascular function. For example, substrates that are removed entirely in a single pass through the pulmonary vasculature might not provide needed sensitivity. In this case, reductions in enzymatic capacity might have to be quite large for effects on intrapulmonary metabolism to be detectable.

Some of the potential pitfalls can be addressed through refinements in techniques. Indeed, if BPAP doses that provide both saturating and nonsaturating concentrations of substrate at enzyme sites in the pulmonary vasculature are measured serially, enzyme kinetics can be determined from the resulting indicator-dilution curves. Thus, the Michaelis constant (K_m) for the enzyme can be calculated, as can A_{max}, which is the product of the maximal velocity (V_{max}) of the reaction and the microvascular plasma volume. From those estimates, information can be obtained on changes in enzyme quality (as measured by affinity) and amount (as measured by V_{max}) in toxicoses or other disease states. With angiotensin-converting enzyme (ACE), for example, a reduction in A_{max} could reflect specific inhibition or destruction of the enzyme or a decrease in capillary surface area. Changes in K_m, however, reflect alterations in endothel-

ial metabolic function that are independent of effects on capillary surface area.

This technique has been used to study effects of pneumotoxicants and ACE inhibitors on pulmonary endothelium. Indeed, alterations in metabolic function of endothelium have been described for such toxicants as PMA and nitrofurantoin and for radiation-induced injury. Studies in rabbits, for example, revealed an increase in K_m for BPAP soon after administration of PMA when no histologic evidence of lung injury was observed (McCormick and Catravas, 1986). The data suggest that, under some circumstances, pulmonary metabolic function can provide a sensitive index of injury to pulmonary endothelium.

As with several other potential markers of lung injury, the use of pulmonary metabolic function to assess endothelial injury in the lung requires further development and validation before it can be considered useful. The equipment and technical sophistication required to perform such assessments are considerable, so modifications would clearly be needed if the method were to be used in routine clinical or screening situations. Thus, it is clear that measurements of pulmonary metabolic functions or other nonbarrier functions of endothelium have not reached the status of clinically useful, diagnostic tests. However, with recent and forthcoming advances in technology, it is not outside the realm of possibility that such techniques will be useful both in the clinic and in the field.

Some of the needs for future research and development are increased basic knowledge of how nonbarrier endothelial functions, such as transporters and enzymes, work in vivo; investigation in animal models of how acute and chronic lung injury changes several endothelial metabolic functions, especially in the absence of surface area phenomena; determination of the specificity and sensitivity of various probes in various injury models with the goal of matching the cause of injury with the probe; and simplification of techniques to make them more useful in human applications.

TABLE 3-3 Summary of Characteristics of Physiologic Assays

Measure	Characteristics[a] and Ratings[b]					
	A	B	C	D	E	F
Respiratory function						
Spirometry	+ +	+	+ +	+ +	+ +	+
Lung mechanics						
Dynamic compliance, resistance, and conductance	+	+	+ +	+	+	+
Oscillation impedance	+-	+	+ +	+ +	+-	+-
Static pressure-volume	+	+-	+	+-	+	+ +
Intrapulmonary distribution						
Single-breath gas washout	+	-	+ +	+	+	+-
Particle distribution						
Exhaled particles	+-	+ +	+ +	+ +	0	+-
Particle deposition	+-	+	+	+-	0	+-
Alveolar-capillary gas transfer						
CO diffusing capacity	+ +	+-	+ +	+-	+-	+
Exercise gas exchange	+ +	+	+ +	+	+	+
Airway reactivity						
Nonspecific reactivity	+ +	+-	+ +	+ +	+	+
Specific reactivity	+	+	+	+-	+ +	+ +
Particle clearance						
Radiolabeled aerosol	+	+	+-	-	+-	+-
Magnetopneumography	-	-	-	+	0	+-
Air-blood barrier function						
Conducting-airway permeability						
Clearance of inhaled DTPA	+-	+	+	-	0	+-
Transepithelial potential	+-	+	+	+-	0	+-
Alveolar permeability by radiolabeled aerosol	+	+	+	-	+-	+-
Vascular permeability						
Radiolabeled protein leakage	+	+	+ +	-	0	+
Chest x ray for edema	+ +	-	+ +	+	-	+-
Extravascular lung water by indicator dilution, PET, or NMR	+	+	+	+-	+-	+-
Rebreathing soluble gases	+	+	+	+-	+ +	+
Endothelial metabolic function	+	+	+	-	+-	+-

[a]Characteristics:

A. Current State of Development. Considerations in this category included the number of groups using the technique, the availability of the required equipment, the magnitude of the present data base, and the degree of standardization of procedures.

B. Estimated Potential for Development. This category reflected the current estimate of the potential for substantial development of the assay beyond its present state. Although it was recognized that advancements are possible for any assay, this category was intended to reflect potential for substantial technical refinements, adaptation for use in large populations, or advancements in ability to interpret results.

C. Current Applicability of Assay to Humans. Primary considerations were the invasiveness of the technique and the requirement for radionuclides. All the assays can be applied to animals, but some are less suitable than others for evaluating humans.

D. Suitability for Measuring Large Numbers of Subjects. The focus of this category was the suitability of the assay for use in studies of large populations of people, as might be required for evaluating effects of some environmental exposures. Considerations included adaptability of equipment for mobile use, length and nature of subject interaction (i.e., degree of cooperation required), resources required to analyze samples and data, and subject safety. For example, a low rating might suggest a low suitability for field use in evaluating hundreds of subjects of various ages and both sexes, whereas the assay might be quite suitable for studies of dozens of selected subjects brought to a stationary facility.

E. Reproducibility. This category focuses on the variability of results within and between subjects.

F. Interpretability. This category reflects the current understanding of (and degree of consensus as to) pathophysiologic correlates, anatomic sites of effect, and causative agents. For many of the assays, there is little disagreement on the physiologic function affected, but the specific mechanism or site of change is uncertain. For example, it is agreed that reduced carbon monoxide diffusing capacity reflects reduced efficiency of alveolar-capillary gas transfer, but the test does not distinguish among the effects of a thickened membrane, reduced surface area, and reduced capillary blood volume.

[b]Ratings:

0 = Unknown, or information is insufficient.

- = Current information suggests inadequate development, little potential for development, little applicability to humans, poor suitability for large populations, poor reproducibility, or poor interpretability.

+- = Current information suggests some development, some potential for development, limited applicability to humans, limited suitability for large populations, questionable reproducibility, or questionable interpretability.

+ = Current information suggests adequate development, potential for further development applicable to humans, suitability for large populations, reproducibility, and interpretability.

+ + = Current information suggests high development or good potential for substantial development, great applicability to humans, great suitability for large populations, reproducibility, or very good interpretability.

SUMMARY

Assays of physiologic function in intact subjects are largely markers of response. Few have potential as indicators of exposure or susceptibility. Measured characteristics often reflect the integrated impact of multiple pathologic alterations; they are seldom indicators of specific, single lesions. The respiratory system responds to injurious agents in only a few ways, so changes in physiologic characteristics are seldom specific to causative agents.

Many assays are well established and have been used extensively for evaluating patients in the clinic and for studying basic physiologic phenomena. In many cases, therefore, there is information on the relationships among changes in pulmonary function values, subjective sense of illness, and performance disability. Although there is much less information on these relationships for some assays, physiologic assays generally provide a key means of estimating the practical meaning of alterations reflected by other types of markers and of estimating the human health impact of environmental exposures. A primary role of the assays, therefore, is to help to determine the extent to which environmental exposures have an impact on health.

We have summarized the current clinical assessment of injury to pulmonary endothelium and described an example of a biologic marker of endothelial cell injury that might become useful in either clinical or screening programs in humans.

Metabolic lung function was chosen as an example of a biologic property under development as a potential marker of lung injury. That choice was intended not to imply that it is expected to be more useful than other potential markers, but rather to illustrate the challenges that must be met in assessing injury to the pulmonary circulation. Indeed, the techniques required to assess this and other biologic markers of endothelial cell injury are cumbersome and require considerable equipment and technical expertise; those are the limitations to their potential application. The need for further validation is also clear. However, the rapid advances in technology that we have witnessed in the recent past and others that are probably imminent might, with commitment and effort, render some of the techniques useful and bring others to light.

Intravascular serotonin is transported into endothelium, where it is either sequestered or metabolized by intracellular monoamine oxidase. The resulting metabolite appears in the pulmonary venous blood. Angiotensin I is hydrolyzed to angiotensin II by angiotensin-converting enzyme on the luminal cell surface. Exposure to endothelium-damaging toxicants might alter these processes of carrier-mediated uptake, metabolism, and sequestration.

Numerous diverse assays are described

in this section, and a tabular summary of their characteristics was thought to be a useful adjunct to the more detailed information in the text. Such tabulation is difficult, because no system of characteristics or rating codes fits all the measurements well. If the difficulty and the resulting cautions are appreciated, however, the information in Table 3-3 can provide a useful overview. The definitions of characteristics and codes follow the table. The definitions of characteristics and codes vary from assay to assay and a plus-minus rating is used because it was not considered appropriate to develop a weighted scoring system leading to a single numeric ranking for each assay. The table is intended as a summary of characteristics.

4

Markers of Altered Structure or Function

WHOLE LUNG

The simplest methods for assessing alterations of lung structure are examination with the naked eye and whole-tissue examination with a dissecting microscope. Of course, access to relevant materials in vivo is problematic.

Events on the pleural surface can be used to establish that lung disease is present (Spencer, 1977). For example, pleural fibrosis and fibrotic adhesions between the visceral and parietal pleurae are clear indicators of exposure to toxic gases or asbestos or of multiple infections (Spencer, 1977). Those alterations can be seen when the chest cavity is opened, and fibrotic pleural plaques can be seen immediately or in wet tissues viewed with dissecting microscopy. The mechanisms that mediate the pathogenesis of pleural fibrogenesis are not at all clear (Bignon et al., 1983). It is conceivable that inflammatory cells provoked to migrate into the pleural cavity produce factors that mediate pleural inflammation and later fibrosis. Thus, analysis of pleural fluids and cells could yield an important marker of impending disease, just as bronchoscopy has provided a window on the cells and fluids of the airways and parenchyma.

The whole lung can be sampled with various techniques to establish the burden of inhaled particles. Lungs collected at autopsy provide exceptionally useful material, because anatomic regions can be sampled and quantitative values derived (Abraham, 1978; Churg and Wright, 1983). Determination of particle types and their correlation with the presence of lung disease have been valuable in increasing the understanding of etiology (Abraham, 1978). Whether such studies will allow for the development of markers of pulmonary disease can be determined only when the nature of the lung burden in occupational and environmental settings is known.

AIRWAYS

A variety of inhaled agents, including oxidant gases and pathogenic microbes, can cause alterations of the large and small airways. The cells of the airways are relatively accessible with bronchoscopy, brushing, and biopsy and therefore have great potential as markers of exposure and injury.

In studying a nasal, tracheal, or bronchial biopsy, it is necessary to establish normal values and appearances of such entities as ciliary beat frequency, cell size, physiologic ion concentration, and structural features as determined by light and electron microscopy. However, the most useful markers are likely to be those related to the basic mechanisms by which airway epithelial cells respond to exposure.

The tracheobronchial lining consists of a pseudostratified epithelium that contains a diverse population of cell types. The ciliated cell is one of the major cell types and is probably a nonproliferative, terminally differentiated cell. The nonciliated population consists of various secretory cells (mucous, Clara, and serous cells, depending on species) and a nonsecretory cell (the basal cell). The Clara, mucous, and basal cells can undergo cell division. Evidence is emerging that the mucous and Clara cells can differentiate into ciliated cells, but the function of the basal cell is not yet established. The cells in the tracheobronchial epithelium can undergo differentiation during vitamin A deficiency and after toxic or mechanical injury. Understanding the mechanisms of differentiation could aid in providing markers of exposure and injury. The availability of biochemical markers of the various differentiated phenotypes is essential because it provides a direct indication of cellular damage. Mucous glycoproteins are used as markers for alterations of mucous cells; specific histochemical staining techniques, as well as biochemical analysis, have been used to identify and characterize secretory cell products (Rearick et al., 1987). For ciliated cells, the presence of dynein appears to be a good chemical marker of structural effects. A low-molecular-weight protein identified in Clara cells appears to be a Clara cell-specific secretory product that can function as a biochemical marker of altered function (Patton et al., 1986). A biochemical marker specific for basal cells has yet to be discovered.

Results of in vivo and in vitro studies indicate that differentiation of tracheobronchial epithelial cells is a multistep process (Jetten et al., 1986) and has several characteristics in common with epidermal differentiation. Like epidermal cells, tracheobronchial epithelial cells undergo cornification. Cornification involves the deposition of a layer of cross-linked protein beneath the plasma membrane. The extensive cross-linking between proteins is catalyzed by the enzyme transglutaminase. Biochemical and immunologic analyses have identified the tracheal transglutaminase as type I (epidermal) transglutaminase. Differentiation of tracheobronchial cells is accompanied by an increase in the activity of transglutaminase by a factor of 20-30 (Jetten and Shirley, 1986). Immunohistochemical staining of tracheas from vitamin A-deficient hamsters with a monoclonal antibody against type I transglutaminase indicated that in vivo synthesis of this enzyme is associated with differentiation of tracheobronchial cells (Jetten and Shirley, 1986).

Squamous cell differentiation is accompanied by an increase in cholesterol sulfate (Rearick and Jetten, 1986). The increase appears to be due to an increase in the enzyme cholesterol sulfotransferase. The high correlation between the expression of that enzyme and increases in cornification indicate that these chemical changes play an important role in differentiation. Such changes could be early indicators of pathologic changes in easily accessible airway lining cells.

Recently, a cDNA library—a library of complementary DNA produced from an RNA template by action of RNA-dependent DNA polymerase (reverse transcriptase)—was established from rabbit tracheal epithelial cells that had undergone squamous cell differentiation (Smits and Jetten, in press). Two cDNAs that are abundant in squamous cells were isolated: SQ 10 and SQ 37, which identify RNAs of 1.0 and 1.25 kb, respectively. The two RNAs appear to be squamous cell-specific and can function as markers for the squamous cell phenotype.

Intermediate filaments have been used in various systems as markers for specific cell types. It has been shown that Clara cells express a keratin profile characteristic of simple epithelium, whereas basal cells express a keratin pattern characteristic of stratified epithelium, such as that consisting of epidermal keratinocytes. Squamous differentiation in vivo, as well as in vitro, is accompanied by qualitative and quantitative changes in keratin expression (Huang et al., 1986). Those findings suggest that keratin expression can be specific for particular cell types of the tracheobronchial epithelium and that changes in keratin expression can indicate particular pathologic alterations, such as metaplasia.

PARENCHYMA

It is in the alveoli that toxic particles and gases exert a fibrogenic influence on the lung interstitial cells that produce connective tissue. The basic cellular and biochemical mechanisms through which inhaled materials cause lung fibrosis are not well established, and it is difficult to ascribe the function of markers to a poorly understood pathologic process. However, animal models of asbestos- and silica-induced disease do offer some potential for elucidation of pathogenesis and thus provide markers of exposure, injury, and disease progression. Detection and analysis of inhaled particles at the alveolar level have been discussed above. The following discusses early cellular responses and methods proposed to study them.

Morphometric Markers of Injury from Exposure to Inhaled Agents

Proliferation of epithelial cells, fibroblasts, and macrophages can be measured in vivo with light and electron microscopy techniques (Evans, 1982). The very earliest responses of the various pulmonary cell types are likely to be proliferative if injury has taken place. For example, both oxidant gases (Evans, 1982) and chrysotile asbestos fibers (Brody and Overby, 1989) cause a rapid incorporation of tritiated thymidine into nuclei of bronchiolar and alveolar cells. That reflects a repair phase, and the degree and extent of the incorporation can be used as a measure of lung injury. Inhaled chrysotile asbestos and instilled crocidolite asbestos cause proliferation of epithelial cells, fibroblasts, and interstitial and alveolar macrophages (Adamson and Bowden,

1986; Brody and Overby, 1989). It is reasonable to speculate that this early proliferative event could serve as a marker of exposure to a toxic gas or particle.

Morphometry is a sensitive quantitative tool that can be of great value in evaluating changes in lung tissue caused by toxic substances. Morphometry is the quantification of three-dimensional structure. Values are derived from an analysis of two-dimensional profiles based on microscopic and other imaging techniques (Weibel, 1979). The advantages of using morphometry to study lung structure include objective measurement of changes in lung structure, identification of subtle changes in tissue structure caused by different agents or by progressive exposure to the same agent, and selective measurement of changes in specific tissue compartments. The enhanced sensitivity of detecting a change in structure by using a morphometric analysis is particularly beneficial when tissue changes due to an experimental treatment are probable, but not obvious. An important application of morphometry involves studies in which subjective grading standards suggest that changes have taken place, but are not conclusive with respect to the relative toxicity of drugs, chemicals, or pollutants at specific exposures or doses. Morphometry eliminates the subjective bias that occurs with many grading techniques used to measure structural changes. It also reduces or eliminates variations in grading between and within observers.

The types of morphometric information that can be obtained in a study of lung parenchyma depend on the resolution used. Light microscopy provides adequate resolution to determine alveolar surface area, proportional volumes of air and tissue, and total number of alveoli. Light microscopy is rapid, is accurate, and can be used with large numbers of specimens. Use of the mean linear intercept (MLI) makes it possible to measure alveolar surface density. MLI measurements have been used to study postnatal lung growth, the normal adult human lung, and lungs with emphysema (Barry and Crapo, 1985). Electron microscopy is better for determining the volumes, thicknesses, and surface densities of specific alveolar compartments. Electron microscopy is required for adequate resolution of alveolar tissue into its components, including type I and type II epithelium, cellular and noncellular interstitium, endothelium, and inflammatory cells. Subcellular components—such as the nucleus, cytoplasm, and cytoplasmic organelles—in specific cell types can be measured with electron microscopy.

Numbers of particles, cells, and subcellular organelles in the alveolar region can be determined morphometrically. The total number of cells in the lung and the distribution of cells among the various major types of alveolar cells have been determined in humans and several species of laboratory animals. The distribution of cell types in the alveolar interstitium has also been estimated (Barry and Crapo, 1985).

The purpose and types of questions to be answered will determine whether light microscopy alone or with electron microscopy is needed. The advantages of light microscopy include rapidity, accuracy, and low cost. Its disadvantages include the need to determine a correction factor for tissue shrinkage due to paraffin embedding of tissues (although other materials that are now available, such as plastic, minimize shrinkage) and restrictions as to the types of alveolar structures that can be easily and reliably measured—airspace, tissue, and capillary volumes; alveolar surface area; mean free path lengths; and numbers of alveoli. Electron microscopy has the advantages of providing quantitative information on specific alveolar compartments and requiring no correction factors, because tissues can be embedded in plastic that allows only negligible shrinkage, as noted above. Its disadvantages include higher costs and requirements of substantially greater labor and time.

Morphometry has been used most commonly to study toxic agents that cause a given kind of injury throughout the entire lung; a random lung sample can be assumed to be representative of the injury. Damage caused by high concentrations of oxygen, a gas that is highly diffusible in lung tissue, is a good example of that type of injury. Some inhaled toxic agents selectively damage the terminal bronchioles and the adjacent alveoli. Those agents include ozone and nitrogen dioxide, which occur as air pollutants in relatively low concentrations, but are more reactive than oxygen. Particles can also selectively injure the small airways in proximal alveolar tissue. The alveoli adjacent to terminal bronchioles are the site of the lesion of centrilobular emphysema, which is at least in part caused by the particles and gases in cigarette smoke.

Rigorous morphometric study of a select region of the lung, such as the terminal bronchioles or the adjacent alveoli, requires sampling techniques that are designed to assess structures that are localized to specific regions of the lung. Common morphometric formulas and basic sampling procedures can be applied, but the methods of tissue selection and data analyses must be adapted. Morphometric studies of the effects of toxic substances on terminal bronchioles and adjacent alveoli have been few, because of the difficulties involved in dealing with samples from specific regions. The appearance of a few studies in recent years directed at the terminal bronchioles or the proximal alveoli indicates increasing awareness of this site of toxic injury and the development of the necessary morphometric techniques (Barry and Crapo, 1985; Chang et al., 1988).

PULMONARY VASCULATURE

To assess the development of structural markers of pulmonary vascular change, one must understand the normal architecture of the arterial and venous circulation. Within the pulmonary circulation are four structural types of pulmonary artery in a progression from hilum to periphery—elastic, muscular, partially muscular, and nonmuscular (Reid, 1979; Meyrick and Reid, 1983). Elastic arteries, by definition, contain more than five elastic laminae, including the internal and external elastic laminae; muscular arteries contain two to five elastic laminae, partially muscular arteries have muscle in only part of the wall; and nonmuscular arteries have no muscle in the wall and are structurally similar to alveolar capillaries (Meyrick and Reid, 1979a). The four types can be related to external diameter: arteries greater than 2,000 μg in external diameter are elastic, those from 150 to 2,000 μm are muscular, and those less than 150 μm are muscular, partially muscular, or nonmuscular.

In addition to the conventional arteries that run and divide with the airways, there is a population of supernumerary arteries (short arterial branches) in the lung that do not (Elliott and Reid, 1965). The supernumerary arteries from the axial pathway are more numerous than the conventional arteries and carry approximately 30% of the blood volume. It is likely, but not certain, that the supernumerary and conventional arteries respond similarly to vascular injury.

Although not described in detail here, the pulmonary veins can be divided into structural regions similar to those of the arteries. In many cases, the pulmonary veins are easily distinguished from pulmonary arteries on the basis of their position in the lung. For example, the walls of the veins are less muscular than those of arteries for a given diameter, and the muscular coat of the bonos is not bounded on its luminal aspect by an internal elastic lamina (Hislop and Reid, 1973).

Two potential outcomes of exposure of the lung to toxic chemicals are progressive restructuring of the lung vasculature and chronic pulmonary hypertension. One of the few well-documented examples of that result in humans is the outbreak of chronic pulmonary hypertension in Europe associated with ingestion of the drug, Aminorex. The paucity of documented examples of chemically induced, chronic vascular injury in the lung other than that incident raises a question as to the importance of the phenomenon in humans. Chronic pulmonary hypertension is difficult to diagnose and might be associated with other maladies, so it could be more common that it seems to be. Indeed, animal studies have revealed that chronic exposure to low doses of several agents that injure endothelium can cause progressive injury to the lung vasculature and pulmonary hypertension.

Moreover, the position of the lung in the circulation renders it the first vascular bed to encounter toxic metabolites of chemicals that are produced by the liver and enter into the venous circulation. The pulmonary vasculature should therefore be considered as a target for chemical insult and chronic injury.

Comprehensive studies of the structure of the pulmonary vessels have used autopsy specimens of pulmonary arteries or veins which were filled with barium gelatin before airway inflation (Reid, 1979; Meyrick and Reid, 1983). The technique allows easy identification of small arteries and veins and full distention of the vascular bed. Full and reproducible distention of the pulmonary circulation allows the use of morphometric techniques; the severity of changes can be assessed, and hypertensive lungs can be compared with normal lungs. The same techniques can be used in lung biopsy tissue (Rabinovitch et al., 1984), although the more subtle and earlier changes of vascular injury might be harder to identify in biopsy tissue.

This section outlines how the pulmonary vasculature responds to changes in hemodynamic behavior, to direct injury, and to environmental changes, such as alterations in ambient oxygen.

Results of pathologic studies of clinical and experimental lung samples have indicated some ways in which the pulmonary vasculature can respond to injury and to changes in pulmonary hemodynamics. Pathologic markers or structural alterations that occur in the pulmonary vessels are usually associated with the development of chronic pulmonary hypertension—e.g., intimal hyperplasia, extension of muscle into smaller and more peripheral arteries than normal, increase in medial thickness of normally muscular arteries, reduction in peripheral arterial volume (seen either as a reduction in number of arteries or as a narrowing of intra-acinar arterial luminal diameter), recanalization of blocked arteries, fibrinoid necrosis, dilatation, and plexiform lesions (Wagenvoort and Wagenvoort, 1977; Harris and Heath, 1986). A single lung biopsy or autopsy specimen might reveal the entire range of morphologic markers of vascular injury or only a few, depending on the stimulus and the severity of the injury. For example, the finding of plexiform lesions or fibrinoid necrosis is associated with end-stage pulmonary vascular disease (Heath and Edwards, 1958; Harris and Heath, 1986); this seemingly restricted response could reflect the paucity of cell types normally encountered in the walls of lung vessels.

In Vivo Observations of Pulmonary Vasculature

Disease of the pulmonary vasculature can be examined in patients both by radiography and by angiography. The radiographic appearance of edema marks pulmonary endothelial injury and is dealt with earlier in this report. Angiographic radiopaque mass marks thrombus formation and vascular occlusion. Rate and abruptness of narrowing and loss of vascular volume can be assessed in pulmonary arterial wedge angiograms; the latter markers have been described in association with chronic pulmonary hypertension secondary to congenital heart defects (Rabinovitch et al., 1981).

Gross Examination of Pulmonary Vasculature

The lung is supplied by two vascular beds: the bronchial, which supplies oxygenated blood to the connective tissue around large arteries, veins, and airways; and the pulmonary, which carries venous blood to the lung capillaries for oxygenation. The pulmonary circulation accounts for more than 95% of the blood volume in the lung. Arterial and venous sides of the pulmonary circulation are easily identified on gross examination of cut lung surfaces, because the arteries run centrally in the acinus (terminal bronchioli and its branches) and the veins at the edge of the acinus. Gross examination of lung slices also allows detection of thrombi, areas of atelectasis, and emphysema, and careful dissection along arterial and venous pathways can provide evidence of congenital defects in the pulmonary vasculature.

Light Microscopic Examination of Pulmonary Arteries

Heath and Edwards (1958) suggested a grading system for the progression of the structural changes of chronic pulmonary hypertension. Grade I represented medial hypertrophy; Grade II, intimal hyperplasia; Grade III, luminal occlusion by intimal hyperplasia; Grade IV, arterial dilatation; Grade V, angiomatoid lesions; Grade VI, fibrinoid necrosis. They considered Grades I-III reversible markers of vascular injury. The grades are readily recognizable in autopsy and biopsy tissue, but only recently have structural markers been correlated with hemodynamic behavior of the lungs.

Rabinovitch and colleagues (1984) refined the grading system of Heath and Edwards, particularly the early reversible changes, by applying morphometric techniques to the lungs of patients with congenital heart defects associated with high flow. Additional structural markers of chronic pulmonary hypertension were identified with that system and graded A, B or C. The quantitative techniques are well described for structural changes in distended and nondistended arteries in autopsy and biopsy tissue (Hislop and Reid, 1973; Reid, 1979; Rabinovitch et al., 1984). Grade A structural changes (extensions of muscle into smaller and more peripheral arteries than normal) are found in patients with an increase in only pulmonary blood flow; Grade B changes (Grade A changes plus increased medial thickness), when both pulmonary arterial pressure and blood flow are increased; and Grade C changes (Grade B changes plus reduction in peripheral arterial volume), when pulmonary vascular resistance is increased.

Those correlations can be extended by examining animal models of chronic pulmonary hypertension: rats exposed to hypoxia and rats given *Crotalaria spectabilis* seeds (which contain monocrotaline). The same structural changes (Grades A, B, and C) can be identified and correlated with hemodynamic behavior (Meyrick and Reid, 1978, 1979b; Rabinovitch et al., 1979; Meyrick et al., 1980). Such studies have shown that the two models of pulmonary hypertension lead to the same structural markers of pulmonary hypertension, but at different rates (Meyrick and Reid, 1983). The structural changes after hypoxia occur faster than those after administration of *Crotalaria*. The severity of each structural marker also differs between the two models (Meyrick and Reid, 1983). For example, loss in peripheral arterial volume after administration of *Crotalaria* is more striking than that after hypoxia, and muscle extension is more severe after hypoxia. Pulmonary arterial pressure also occurs faster after hypoxia. Correlation of the structural changes with the hemodynamic behavior in both models reveals that the best structural correlate of increased pulmonary arterial pressure is extension of muscle into small arteries; for *Crotalaria* administration, it is increased medial thickness.

The data also suggest that the speed of the changes can depend on the stimulus for the development of chronic pulmonary hypertension. With hypoxia, one of the mechanisms involved is almost certainly hypoxic vasoconstriction; with *Crotalaria* administration, the changes are thought to be secondary to endothelial damage, and the role of vasoconstriction is less certain. Additional evidence that endothelial damage can contribute to the development of pulmonary hypertension was provided by Jones and colleagues (Jones et al., 1984) who used rats exposed to hyperoxic conditions. Recent data on sheep given repeated infusions of endotoxin (Meyrick and Brigham, 1986) and sheep subjected to continuous air embolization (Meyrick et al., 1987; Perkett et al., 1988) suggest that inflammation of the lung can be a trigger. The data suggest that there are several stimuli of development of chronic pulmonary hypertension and that the structural markers of pulmonary hypertension might not be identical in each type of hypertension. Development of chronic pulmonary hypertension is likely to be seen in association with hypoxia, hyperoxia, prolonged vasoconstriction, chronic or repeated lung inflammation, and endothelial injury.

The variation in severity and appearance of structural markers of chronic pulmonary

hypertension and the range of initiating stimuli are borne out in the clinical setting (Reid, 1979; Meyrick and Reid, 1983). For example, natives of high-altitude environments and children with persistent fetal circulation (pulmonary hypertension following right-to-left blood shunting) have marked muscle extension, and patients with chronic bronchitis and emphysema, cystic fibrosis, and primary pulmonary hypertension have a less severe change. Reduction in peripheral arterial volume is striking in patients with primary pulmonary hypertension, but blood volume is within normal limits in cases of persistent pulmonary hypertension. Plexiform lesions are found in patients with primary pulmonary hypertension and in older patients with congenital heart disease, but have not been seen in the hypoxic forms of chronic pulmonary hypertension.

Not only the pulmonary arterial circulation is limited in its response to injury, but also the veins. On the venous side of the circulation, "arterialization" of the veins occurs if pressure in the veins is chronically increased. The structural markers are a medial coat and an internal elastic lamina in the venous walls. Those changes are often accompanied by intimal fibrosis (Harris and Heath, 1986). An increase in pressure also leads to the development of the structural markers of pulmonary hypertension on the arterial side of the circulation.

Electron Microscopic Examination of Pulmonary Arteries

Electron microscopy shows the major cell types in the walls of the pulmonary circulation to be endothelial cells, smooth muscle cells and their precursors and fibroblasts. In experimental models of chronic pulmonary hypertension, such as the rats exposed to hypoxia (Meyrick and Reid, 1978, 1979b) or given *Crotalaria spectabilis* (Meyrick et al., 1980), it has been shown that the precursor smooth muscle cells undergo hypertrophy and, at least in hypoxic animals, cell division; that accounts for the appearance of muscle in the normally nonmuscular intra-acinar pulmonary arteries (Meyrick and Reid,

1983). Encroachment of the new muscle cells and endothelial hypertrophy contribute to the reduction in peripheral arterial volume (Meyrick and Reid, 1983). Ultrastructural studies also show that smooth muscle cell hypertrophy and alterations in connective tissue synthesis and accumulation contribute to the increased medial thickness seen in preacinar arteries (Meyrick and Reid, 1980).

Future Directions

Few reports have suggested alterations in bloodborne mediators in patients with chronic pulmonary hypertension although Geggel and associates have suggested that measurements are possible. Increases in ristocetin cofactor activity relative to plasma von Willebrand factor have been reported in patients with primary pulmonary hypertension (Geggel et al., 1987). Similarly, patients with primary pulmonary hypertension and monocrotaline-treated rats with increased pulmonary arterial pressure have increased plasma copper concentrations (Ganey and Roth, 1987). The origin of the latter change is unclear, but such observations deserve further attention with regard to development of markers of chronic pulmonary vascular changes. Advancements in this direction will be of obvious importance to patients, especially if changes can be detected early in the development of the disease.

Results of recent biochemical and molecular studies of chronic pulmonary hypertension have indicated that the development of the structural changes might depend on the stimulus. For example, in the monocrotaline model of pulmonary hypertension in rats, an increase in elastase activity in the preacinar arteries resulted in thinning and fragmentation of the internal elastic lamina of the thickened walls (Todorovich et al., 1986). In the hypoxic neonatal cow, the increased thickness of the walls of the preacinar arteries included proliferation of adventitial fibroblasts invoked by a growth factor released from the medial muscle cells (Mecham et al., 1987). In hypoxic rats, treatment with a latherogen (beta-aminopropionitrile) or with an inhibitor of collagen production (cis-4-hydroxy-L-proline) caused attenuation of the structural markers of pulmonary hypertension examined, as well as partial protection against the increase in pulmonary arterial pressure (Kerr et al., 1984, 1987). Thus, markers of pulmonary vascular disease are likely to be diverse and to depend on the initial stimulus of hypertension. The introduction of such techniques as in situ hybridization is likely to advance our understanding of these alterations at the cellular and molecular level and to yield markers useful in the early detection of chronic pulmonary hypertension.

5

Markers of Inflammatory and Immune Response

Inflammation in the respiratory tract can be caused by injury, immune response, or infection. All three routes of induction of inflammation are of interest in studying the health effects of respirable pollutants. Inflammation due to injury from inhaled toxicants is a measure of the cytotoxicity of the pollutants; markers of the progress of the inflammation could potentially be used in a predictive fashion to detect the early stages of irreversible structural changes in the lung, such as fibrosis and emphysema. Many inhaled toxicants—such as isocyanates, cotton dust, and beryllium—induce an immune response that constitutes the major adverse health effect of exposure. The inflammatory response to infectious agents is not the topic of this report, but is of interest in relation to the potential for some inhaled pollutants to reduce the ability of the body to resist infection.

In this chapter, we will discuss markers to detect and require the inflammation induced by all three types of agents. First is a discussion of markers of injury induced in the respiratory tract by inhaled toxic materials. That is followed by a discussion of the effect of pollutants on the infectivity of pathogens in the respiratory tract. Finally, a major portion of the chapter is devoted to a discussion of the immune response of the lung, markers of this type of response and the use of memory cells of the immune system as markers of both exposure and adverse health effects in the respiratory tract.

INFLAMMATORY RESPONSE TO INHALED TOXINS

Epithelial cells and resident macrophages in the respiratory tract are the points of first contact of the body with inhaled toxicants. The ensuing injury or death of those cells induces an inflammatory response characterized by the release of cytoplasmic enzymes from the damaged or lysed cells and the recruitment of neutrophils to the site of injury. An increase in the permeability of the alveolar-capillary barrier is accompanied by the transudation of serum protein. Macrophages may increase in number; if the toxic material is a particle, there will be a release of hydrolytic enzymes from the macrophages, either during phagocytosis, after lysis of the macrophage, or as an active secretory process.

The inflammatory process provides many markers that can be used to detect and measure the response and to follow its progress toward resolution or chronic inflammation. Lactate dehydrogenase, a cytoplasmic enzyme released from damaged or lysed cells, can be used as a marker of cytotoxicity. An increase in this enzyme can be used to distinguish between toxic events and physiologic responses. The presence of neutrophils or increased serum protein in the epithelial lining fluid can be used as a marker of the inflammation induced by the injury. The activity of hydrolytic and proteolytic enzymes released by the phagocytic cells has been shown to correlate with the toxicity of particles (Beck et al., 1982; Henderson, 1988a,b).

The advent of fiberoptic bronchoscopy has allowed sampling of fluids lining the respiratory tract for analyses of the above markers. The use of the technique is described fully in Chapter 6. The analysis of epithelial lining fluid (ELF) for markers of inflammation has some advantages over older methods of detecting inflammation. First, a pulmonary inflammatory response can be detected with analysis of the bronchoalveolar fluids before it can be detected with radiography (Fahey et al., 1982). In addition, the use of the markers in the lining fluids allows measurement of the degree of inflammation, which is useful not only for determining the progress of an inflammatory response, but also for ranking inhaled compounds for toxicity in animal studies. Inflammatory responses in the nose and in the upper respiratory tract can be detected by site-specific sampling of the lining fluids. The biochemical and cellular content of the ELF can also provide information on the type of inflammatory response and the stage of the disease process. Several examples of the use of ELF analysis for the detection of toxicant-induced respiratory tract inflammation are given in Chapter 6. The major uses of the markers has been in animal toxicity studies to rank a series of compounds for toxicity and to study the mechanisms of toxicant-induced lung disease. Studies in this field have been reviewed (Henderson, 1988a,b).

INFLAMMATORY RESPONSE TO MICROBIAL INFECTIONS

The respiratory tract constitutes the primary mammalian portal of entry for many pathogens. For some microbial infections, the bronchopulmonary mucosa serves as a benign substrate for initial replication events that lead to eventual systemic spread without producing any clinical disease locally. For other infectious agents, however, the respiratory tract is the principal target for the disease-producing potential. It is estimated that respiratory viruses are responsible for 5-6% of all deaths and about 60% of deaths related to respiratory disease (Ogra et al., 1984).

Ample evidence supports the conclusion that normal AMs can ingest and kill many types of microorganisms that gain entrance to the respiratory tract (Jakab, 1984). Some microorganisms, particularly the virulent intracellular parasites, can survive in normal macrophages; it is apparent that this class of parasite can be controlled only when the forces of acquired immunity orchestrate the macrophage system into antimicrobial action that is more potent, both qualitatively and quantitatively.

Various chemicals in the environment and workplace affect the immune response, as determined by one or more of the many tests available to measure various components of the immune system (Faith et al., 1980; Sharma, 1981; Luster et al., 1988). Extensive evidence from animal lung infectivity models points to the detrimental effects of air pollution on various defense mechanisms of the lung (Neiman et al., 1977; Fauci et al., 1984). Results of epidemiologic studies indicate that living in urban areas increases the incidence of airway infections (Hong, 1976; Penn, 1978), although direct correlations of infection with specific pollutants have not been found. In vivo and in vitro studies, using mainly animal models, have shown that ozone impairs AM phagocytic function (U.S. Environmental Protection Agency, 1984). The success of AMs in digesting organic particles and organisms depends on lysosomal hydrolytic enzymes, including acid phosphatase, cathepsins, lysozyme, beta-glucuronidase, beta-galactosidase, arylsulfatase, and beta-glucosaminidase. Exposure to ozone in vivo and in vitro depressed the intracellular activity of lysozyme, beta-glucuronidase, and acid phosphatase from rabbit AMs (Hurst et al., 1970).

Kimura and Goldstein (1981) demonstrated a decrease in the bactericidal enzyme lysozyme in AMs from rabbits exposed to ozone at 0.25 ppm. That finding paralleled the increased susceptibility of those animals to various infectious agents (Coffin et al., 1968; Ehrlich et al., 1977; Miller et al., 1978). Other products of the AMs that are important in phagocytosis and have antibacterial activity are superoxide anions (O_2^-), hydroxyl radical ($OH^•$), and H_2O_2. In vivo exposure of rats to ozone at 0.9–3.2 ppm resulted in a dose-dependent decrease in O_2^- production by AMs from the exposed animals (Amoruso et al., 1981; Witz et al., 1983). Thus, the studies in animal models have shown that ozone depresses many components of AM function that are important in phagocytosis and in protecting the lung from microorganisms or particles present in the environment and that can be measured and therefore can serve as markers of exposure.

In vitro studies can also be used to improve understanding of in vivo phenomena. Phagocytosis is a well-organized process, made up of several integrated steps, many of which are adversely affected by ozone. McAllen et al. (1981) have noted that AMs obtained from rats after in vivo exposure to ozone at 1 ppm exhibit decreased mobility. AMs in lavage fluid from the lungs of rodents that had been briefly exposed to ozone at 0.5–1.0 ppm display a lower ability to engulf bacteria than AMs from control animals. However, if the animals are exposed to ozone at 0.8 ppm for a longer period before lung lavage, the ability to incorporate carbon-coated latex microspheres of isolated AMs is increased. The data suggest that phagocytic activity is impaired soon after exposure to ozone, but that AM function recovers if the insult persists. Whether those changes are due to the influx of unexposed AMs into the lung or to adaptation by resident AMs is not clear.

Several investigators have recently started to examine the effect of inhaled pollutants on human subjects' ability to resist infections. Frampton et al. (1987) exposed normal human subjects to NO_2 in controlled chamber conditions. AMs were obtained from the subjects by BAL 3.5 hours after exposure and exposed in vitro to influenza virus. It was observed that the AMs obtained from subjects exposed to NO_2 at 0.6 ppm continuously (3.5 hours) were able to inactivate the virus significantly less than those obtained from subjects exposed to clean air. No major changes in the cell numbers in the BAL fluid from the exposed subjects were observed, but several biochemical changes indicated that NO_2 did induce inflammation in the exposed subjects. In another study, Kulle et al. (1987), exposed normal human subjects to NO_2 at different concentrations, and then a live attenuated cold-adapted influenza A virus was administered intranasally to all subjects. Infection was determined by virus recovery and a 4-fold or greater increase in antibody titer. The results suggest that subjects that were exposed to NO_2 at 1 or 2 ppm for 2 hours/day for 3 days and inhaled the virus on day 2 had small, reproducible signs of infectivity. The approaches taken by the different investigators were different—Frampton et al. exposed subjects to the pollutant in vivo and studied infectivity in vitro, and Kulle et al. exposed subjects to both the pollutant and the infectious agent—but the results collectively suggest a decrease in host defense against the viral infection. In both studies, spirometrically measured pulmonary functions were unchanged by exposure to the pollutants.

IMMUNE RESPONSE

The human respiratory tract contains a complex array of host defenses—anatomic barriers, mucociliary clearance, phagocytic cells, and various components of cellular and humoral immunity—that collectively cleanse inhaled air and inactivate infectious and other injurious agents that are inhaled (Reynolds, 1979; Reynolds and Merrill, 1981). In particular, the mucosal lining of the small airways and alveolar airspaces contains many components of the immune system that are important in providing protection of the normal lung. However, some of the components also play an important role in immunologic lung disease. This section reviews the general features of the immune system and how it operates in the lung.

Antigen-antibody complexes are the basis of immune response. The afferent phase of the immune response usually begins with antigen processing by phagocytes, such as macrophages. That includes degradation of foreign substances and exposure of lymphocytes to antigens, which stimulates the production of antibody, sensitized cells, or both (Figure 5-1). Interactions of macrophages stimulated by antigens with cells in lymphoid tissue result predominantly in a cellular or humoral immune response. Cellular immune responses (delayed hypersensitivity) are mediated by thymus-derived lymphocytes—T cells. Antigen interaction with T cells usually leads to their proliferation. It is now recognized that T-cell proliferation and the generation of effector cells occur separately. Antigens interact with macrophages, and interleukin-1 (Il-1) stimulates resting T cells. The T cells then can respond to a growth factor called interleukin-2 (Il-2). Among the progeny of antigen-stimulated T cells are memory cells, which

respond quickly to later challenge with the original antigen; killer cells (or natural killer cells, NK cells), which destroy alien cells; and effector T cells, which produce molecules called lymphokines. Il-2 signals T cells to produce more T cells and effector T cells. Lymphokines can play an important role in the generation of an inflammatory response, particularly one involving cell-mediated immunity. For example, initiation and development of granulomas are thought to arise from the secretion of lymphokines that influence macrophage motility, activation, and function.

Humoral responses are the end result of antigen interaction with marrow-derived or bursal-cell-equivalent lymphocytes (B cells). B-cell function is regulated by at least two subpopulations of T cells: helper T cells (Th cells) are required for optimal production of antibody to most antigens, and suppressor T cells (Ts cells) are required for inhibition or modulation of the humoral response once it is initiated.

T-cell subsets in humans have been shown to express distinct differentiation antigens, which can be identified with monoclonal antibodies to T cells.

At each step in the sequence of immune stimulation, immunocompetent cells are activated and liberate soluble mediators. For example, when activated by immune stimulation, macrophages can produce various potentially injurious agents, including arachidonic acid metabolites,

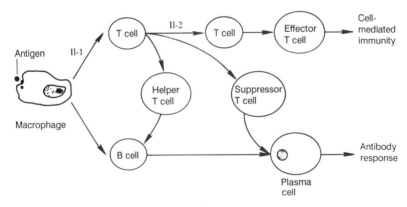

FIGURE 5-1 Cellular interactions involved in generation of immune response. Antigen presentation leads to stimulation of T-cell or B-cell systems. Factors involved in T-cell system include interleukin-1, which stimulates T cells to acquire receptor for T-cell growth factor called interleukin-2 (Il-2); same subpopulation of T cells can also secrete Il-2.

free oxygen radicals, and growth factors capable of initiating abnormal growth and metabolism of fibroblasts. Similarly, activated lymphocytes can produce soluble mediators that can act on immunocompetent cells and other cells, including interferon-gamma capable of activating macrophages and chemotactic factors capable of activating other phagocytic cells, particularly neutrophils. Thus, although immune stimulation is initiated by a specific antigen, secondary inflammatory mediators can appear as the cascade of cells and inflammatory signals progresses. Many of the cells, particularly macrophages, can be stimulated by nonspecific agents, and the appearance of soluble mediators does not necessarily indicate that an immune response has been initiated. The potential use of some of the inflammatory mediators as specific biologic markers of injury must be viewed in light of the fact that they can be initiated by both specific and nonspecific stimuli.

The pulmonary immune cells are heterogeneous, and external stimuli, such as pollutants, can modulate their behavior both qualitatively and quantitatively. The changes can be used to assess effects of exposure. Characteristics that can be monitored, such as cell numbers and cell-surface markers, can be considered biologic markers. Some of the changes can be transient (reversible); others can be chronic or irreversible. Examples of modulation in the immune cells in the lung include changes in the proportions of subpopulations of immune cells (e.g., the TH/TS ratio), in the extent of expression (density) of cell-surface markers, in the cytolytic capacity of T cells or natural killer cells, and in the ability of phagocytes to ingest particles.

Various components of lung fluid that are associated with the inflammatory response can serve as markers. The fiberoptic bronchoscope has made access to the trachea and major airways routine, and this versatile instrument has contributed enormously to the care and diagnosis of patients with lung diseases (Sackner et al., 1972). Since the late 1960s, BAL has been incorporated into fiberoptic bronchoscopy and has proved to be safe and reliable for sampling airway and alveolar fluid and cellular components in normal lungs (Reynolds and Newball, 1974) and diseased lungs (Reynolds et al., 1977; Crystal et al., 1981). BAL has made possible extensive study of the pathogenic roles played by immune and inflammatory reactions in the respiratory tract and important advances in understanding the pathogenesis of various forms of obstructive, inflammatory, and interstitial lung disease. Much of the information reviewed in this chapter has been obtained with BAL (Reynolds, 1987).

In lavage of the lungs of a normal nonsmoker, approximately 10-15 million respiratory cells typically are recovered. The number of cells obtained in BAL can vary widely. For instance, the number of cells is increased by a factor of approximately 4-5 in lavage fluid from a cigarette-smoking patient and is correlated roughly with the intensity of smoking. Approximately 90% of the cells in lavage fluid are alveolar macrophages (AMs). Most of the remaining cells are lymphocytes, and usually only a few polymorphonuclear leukocytes (PMNs) are found. Eosinophils and basophils are rarely detected. In smokers, the cell yield far exceeds that of nonsmokers. Therefore, although the percentage of lymphocytes is diminished, the absolute number is actually increased. Analysis of the differential counts and more recently the use of monoclonal antibodies and flow cytometry of the cells in BAL specimens have been found to provide useful markers of a variety of pulmonary disorders, particularly the granulomatous and nongranulomatous interstitial lung diseases, and of effects of exposure to pollutants.

AMs are the principal phagocytic cell in the airways and seem to play a pivotal role in initiating and modulating the pulmonary immune response (Merrill et al., 1982). Phagocytosis of microorganisms and other foreign particles in the alveoli by macrophages is an important defense mechanism of the lung against infection and other forms of external assault by inhalation. AMs interact with other cells and foreign material in the lung via membrane receptors. Surface receptors for the Fc portion of IgG and for complement fragments C3b and C3d have been identified (Reynolds et al., 1975). The activity of the phagocytic system generally depends on the specific immunoglobulins IgA and IgG—IgA probably functions mainly as an antitoxin and in the neutralization of viral infectivity, whereas IgG is also important in promoting phagocytosis.

Although complement concentrations are low in the airways, complement can be an important participant in infections when the inflammatory response promotes transduction of complement, as well as of serum immunoglobulins. Once phagocytosis is completed, killing of an ingested organism is mediated by phagosome-lysosome fusion, degranulation, and the elaboration of digestive enzymes and toxic oxygen species. Present evidence suggests that AMs use metabolically generated H_2O_2 in conjunction with some type of oxidase to kill microorganisms. The hydrolases in the lysosomes probably play a major role in digesting a phagocytosed microbial carcass. It is also likely that mobilized and activated lymphocytes and macrophages can produce an exterior milieu that is adverse for at least some microorganisms. The various specific immunologic mechanisms probably act in concert with nonspecific factors in mucus, such as lysozyme and other nonantibody antimicrobial agents; some events of inflammation, as well as the mucociliary approaches, also are important contributions to the overall defense of the lung.

AMs participate further in the regulation of the inflammatory and immune process in the lungs by secreting a variety of soluble mediators, including products of the arachidonic acid pathway, which seem to play an important role in inflammation (Hunninghake et al., 1980a; Slauson, 1982).

AMs secrete chemotactic factors for neutrophils that cause influx of these cells into the lung parenchyma and alveolar space, where they can participate actively in the inflammatory response. AMs have a wide range of other secretory capabilities and have been shown to secrete such products as colony-stimulating factor, superoxide anions, and various enzymes, including collagenase, neutral protease, and elastase. Among the most potent inflammatory substances produced by AMs are products of arachidonic acid (Slauson, 1982). Some of the better-known mediators of the inflammatory response include prostaglandin F_2 (PGF_2 alpha) and the chemotactic factor LTB_4. Those arachidonic mediators and their occurrence are described below. AMs also play a unique role in the development of an immune response to a novel antigen by presenting bound or ingested antigen to T lymphocytes. About 8-10% of the cells in the BAL fluid recovered from normal human lungs are lymphocytes. As in blood, three subpopulations of lung lymphocytes (T, B, and NK cells) can be discerned on the basis of their differing surface markers. The proportions of T and B lymphocytes in lavage fluid from the airway lumina have been found to approximate closely those in peripheral blood (Hunninghake and Crystal, 1981a). Whereas T lymphocytes make up about 60-70% of the lymphocytes found in normal BAL fluid, only about 10-15% of the lymphocytes have surface immunoglobulin and can be identified as B cells. A portion of the lymphocyte population that cannot be classified by classical T- and B-cell markers has been identified as NK cells.

PMNs are not usually found in large numbers in BAL fluid from normal lungs (Reynolds, 1987). In lavage fluid from the lungs of smokers (Young and Reynolds, 1984) or subjects that were exposed to ozone (Seltzer et al., 1986; Koren et al., 1989), an increased percentage and greatly increased absolute numbers of neutrophils are found. As participants in the inflammatory response in lung tissue, PMNs migrate into the lung from the blood under the influence of one or several chemotactic factors, including complement component C5a and soluble factors secreted by AMs. They might be considered as a secondary line of phagocytic defense of the lungs, which can be recruited into the airspaces in response to exposure to microbial agents or other inhaled materials. Some of the observed changes have been shown to have a predictive value; others represent more progressive biologic changes.

Acquired Local Immune Response

In defining the role of the lungs' immune system in the generation of specific biologic markers of injury or disease, it is important to consider the central feature of an immunologic response—that it is a specific reaction to an antigenic stimulus and is capable of distinguishing proteins that differ by only a few amino acids. A substance that is inhaled and can act as a unique and foreign antigen elicits an immune response that is manifested predominantly in the form of antibodies or sensitized cells. (This manifestation is the cellular immune response.) As discussed below, local immune responses in the lung operate in concert with other mechanisms (such as mucociliary clearance) in recognizing, transporting, and eliminating inhaled foreign agents. Obviously, local immune responses are very important; their failure can result in injury and tissue damage in the lung.

With respect to local immune system as a generator of markers, two underlying possibilities or conditions need to be considered: the presence of antigen-specific antibodies or cells indicates that the host has been exposed to an antigen at some time, even if no longer harboring it; and specific immune responses might indicate the presence and persistence of an antigen that produces a chronic inflammatory response that leads to tissue injury. Thus, the products of the immune system can be used as markers of a host's exposure to an antigen that might or might not be responsible for tissue injury and disease.

In an operational model, the lung's immune system can be viewed as having three distinct compartments, each containing immunocompetent cells (lymphocytes and macrophages): the bronchoalveolar airspaces, the submucosal or secretory antibody system lying beneath the lamina propria of the tracheobronchial tree, and a network of lymphatic vessels and lymph nodes lining the tracheobronchial tree (Daniele, 1980). In each of these compartments, the potential exists for lymphocyte-macrophage interaction and the generation of immune responses.

In the last 5 years, most of our knowledge about cell-mediated immune responses in the lung has come from studies involving cells recovered from bronchoalveolar airspaces with lavage in humans or in experimental animals (Daniele et al., 1985).

Until the advent of the flexible fiberoptic bronchoscope, little was known about the cells and secretions in the bronchoalveolar airspaces of the human lung. It has since been observed that, in nonsmoking adults, cell yields equal 10-15 x 10^6 cells/100 ml of lavage fluid (Dauber et al., 1979); AMs are the predominant cell type (80-90%), lymphocytes constitute about 10% of the cells (Reynolds et al., 1977; Dauber et al., 1979), and neutrophils, eosinophils, and basophils constitute less than 1% of the cells. In smokers, the cell yield is some 4 times as great; macrophages usually account for 90% or more of recovered cells, and lymphocytes for 1-5% (Daniele et al., 1977b; Reynolds et al., 1977; Dauber et al., 1979); there can also be a slightly higher proportion of neutrophils (1-4%).

The distribution of lymphocyte subpopulations in the lavage fluid is similar to that in blood, with T cells accounting for 60-70% of the lymphocytes and B cells 5-10%. The ratio of Th cells to Ts cells is 1.6:1 (Daniele et al., 1975; Dauber et al., 1979; Hunninghake et al., 1979b; Hunninghake and Crystal, 1981a).

The major soluble constituents in lavage fluid are IgG and IgA; their concentrations reflect rates of active transport across the bronchial epithelium (Reynolds et al., 1977; Low et al., 1978; Hunninghake et al., 1979b). Little or no IgM is present. Components of both the classical and alternative pathways have been identified in lavage fluid, but C5 appears to be absent (Robins et al., 1982). Other inflammatory derivatives have been detected in lavage fluid, including alpha$_1$-antitrypsin.

Much needs to be learned about the role of lymphocytes in the normal human lung, particularly with respect to their initiation and development of immune responses. The evidence is more substantial in experimental animals (Daniele, 1980). Lymphocytes recovered with lung lavage from guinea pigs and rabbits respond to antigens introduced into the respiratory tract by producing antibody and lymphokines (such as macrophage-migration inhibition factor, MIF). Furthermore, lung lymphocytes can demonstrate an anamnestic response to airborne antigens and, depending on the type and dose of antigen, exhibit a capacity to respond that is independent of systemic lymphoid tissue. Thus, results of animal experiments indicate that localized immune responses can occur in the bronchoalveolar airspaces.

Alternatively, it has been proposed that the cells and secretions in the bronchoalveolar airspaces are deployed so as to prevent entry of antigenic particles beyond the mucosal barrier and to deter antigen interaction with organized lung lymphoid tissue. According to the notion of "immune exclusion," the primary function of AMs is to ingest particles and remove them from the lung, rather than to transport them to submucosal and tracheobronchial lymph nodes, where lymphocytes and tissue macrophages might interact. Which of the two hypotheses is correct remains to be settled.

The two hypotheses might not be mutually exclusive. The nonspecific activities of AMs and the mucociliary blanket might be entirely adequate for expelling some inhaled inert substances. Nonspecific clearance mechanisms would not suffice for other antigens, such as microorganisms with capsular membranes that resist phagocytosis, and the aid of specific antibody and cells in bronchial secretions would be required for effective phagocytosis, killing, and clearance. The generation of a specific immune response consisting of either antibody or cells in bronchial secretions requires, however, that the inciting antigen in some way penetrate the mucosal barrier and stimulate submucosal lymphoid cells. That condition is also required for any inhaled particles (e.g., allergens and organic particles) that result in local humoral and cellular immune responses. It should also be emphasized that initially only a relatively small fraction of the inhaled antigenic load might be required for stimulating submucosal lymphoid tissue. Once initiated, the secretion of antibody or the appearance of sensitized cells in the airspaces would greatly increase the exclusion of the same or similar inhaled antigens on later challenges.

The degree to which nonspecific defenses interact with specific immune responses in the lung remains ill defined. It probably depends on the size of the particle, the antigenic load, and the physicochemical characteristics of the particle, which are related to its antigenicity, toxicity, and, perhaps most important, biologic properties (e.g., type of virus and capsulated bacteria).

Those are some of the variables that determine whether inhaled particles and microorganisms are contained or eliminated or result in lung injury and disease. Perhaps equally or more important are the unique genetic properties of the host, especially the immune responses that are linked and controlled by the immune-response genes. The latter consideration is particularly relevant for two immunologic diseases, hypersensitivity pneumonitis and chronic berylliosis, that are discussed below. In both, only a minority of persons equally exposed to the airborne agents develop disease.

Examination of cells and secretions in BAL fluid from patients with immunologic lung diseases has provided important insights into pathogenesis.

First, the lung can be the site of a compartmentalized inflammatory response (Daniele, 1980), as in hypersensitivity pneumonitis, in which the disease is restricted to the lung. In other systemic disorders, the inflammatory response that evolves in the lung might not be reflected in the peripheral blood (Daniele et al., 1980). The reason for the difference is unclear; one hypothesis is that the lung, when it is involved, acts as a selective target for acute (neutrophils) or chronic (lymphocytes and monocytes) inflammatory cells, which are increased in the pulmonary parenchyma as well as in the lavage fluid (Crystal et al., 1981).

Second, pulmonary lavage has established the existence of two predominant types of chronic inflammatory response in the lung, one involving neutrophils and macrophages (idiopathic pulmonary fibrosis) and the other involving lymphocytes and macrophages (hypersensitivity pneumonitis and berylliosis).

Finally, several laboratories have found in studies of pneumonitis a heightened state of activation of these inflammatory cells (Daniele et al., 1980; Crystal et al., 1981). Lymphocyte activation probably reflects immune stimulation in cases of hypersensitivity pneumonitis and berylliosis.

In summary, the ability to detect sensitized cells or antibodies that are specifically reactive to large complex organic antigens (as in hypersensitivity pneumonitis) or simple elements that behave as haptens (as in berylliosis) can serve as a useful paradigm for investigating other inhalational diseases in which an immunologic response is predominant in pathogenesis. The presence of specific responses in the lung indicates that the subject has been exposed to a foreign antigen; it does not necessarily mean that the antigen is causing disease. For example, in both hypersensitivity pneumonitis and chronic berylliosis, it is still unclear whether the presence of sensitized cells or antibodies in BAL fluid indicates that a patient has or will have disease related to the foreign substances found.

Acquired Antigen-Specific Immune Response

In Vivo Challenge

Testing for immune response has often included testing of whole animals or humans. With such testing, the interaction of several components of the immune system can be tested at once, and actual body response can be measured, so that one need not rely on extrapolation from results obtained in vitro. However, in vivo challenge has several difficulties: the risk of a serious adverse reaction, including anaphylaxis; the difficulty of separating a nonspecific from a specific response; the difficulty of interpreting whether a response in one area reflects a response in another area; and the difficulty of purifying an antigen to be specific enough for testing and suitable for administration without causing nonspecific damage.

The immune response has been divided into four groups summarized in Table 5-1 (Bellanti, 1985). Skin testing usually elicits Type I reactions, although Type IV reactions can be detected in skin. Cell-mediated immunity is tested by examining the skin site 24-48 hours after injection; this can be done to determine whether a subject has been infected with tuberculosis—as with the PPD skin test (Snider, 1982)—or to determine whether a patient is anergic (not reacting to any of the common antigens, such as those of tetanus or mumps). Testing for granuloma formation can use the Kveim antigen (sarcoid tissue antigen) (Chase, 1961). Skin testing for delayed reactions has not been routinely used for detecting sensitivity to pulmonary toxicants.

With further sophistication, a challenge might be graded not only by the amount of visible inflammation present, but by other factors, such as the influx of inflammatory cells and the presence of inflammatory mediators, including histamine, immunoglobulins, and immune complexes.

Skin Testing

A standard method for testing for reaction to a possible pollutant is skin test-

TABLE 5-1 Immunologic Mechanisms of Tissue Injury

Type	Manifestations	Mediators
I	Immediate hypersensitivity reactions	IgE and other immunoglobulins
II	Antibody-directed reactions	IgG and IgM
III	Formation of antigen-antibody complexes	Mainly IgG
IV	Delayed hypersensitivity (cell-mediated) reactions	Sensitized T lymphocytes

ing. Allergists have used skin testing extensively to identify substances to which a patient is allergic (Norman, 1980). Its potential use in environmental studies in toxicology is based mostly on its simplicity of application and interpretability. The procedure is relatively safe, although some subjects are allergic to antigens and anaphylaxis has been reported after skin testing in a few subjects. Usual precautions in skin testing include testing with the lowest possible dose and observing subjects for some time after testing.

Methods of skin testing include prick testing, scratch testing, and intradermal injection, in order of increasing dose. The skin prick test is the safest, in that a very small amount of antigen is injected.

Reactions to a prick test are read at 10 minutes; a reaction indicates an immediate type of sensitivity. Patients with dermatographism will have a false-positive wheal; otherwise, the test is readily assessed. Pepys's group has used the prick test for many years to evaluate exposure of platinum workers (Cleare et al., 1976). Platinum salts can be highly reactive, and systemic reactions can occur even to scratch tests, which therefore were considered too risky for general surveillance testing. Nevertheless, the authors also found that the prick test was better for differentiating between controls and reactive subjects.

A major difficulty with skin testing has involved the preparation of a sufficiently reactive antigen. Most substances studied are haptens and become antigenic on combination with high-molecular-weight carriers. That can happen at the injection site. For example, phthalyl acid anhydride is an essential reagent in the manufacture of epoxy resins and some paints. It is highly reactive, and skin testing can be performed directly. Positive skin tests have been seen in documented cases of asthma induced by phthalyl anhydride (Maccia et al., 1976). Most substances do not induce responses by themselves and have to be conjugated to human serum albumin before testing. Some have been found to be reliable skin-test reagents for particular environmental pollutants (Zeiss et al., 1983).

Another difficulty with skin testing has involved the need to relate the findings with the pulmonary symptoms of the subjects. A positive skin prick test in platinum refiners is a more specific and sensitive index of disease than are some clinical symptoms (Dally et al., 1980). For example, skin prick tests of mouse and rat urine extracts in laboratory-animal workers have yielded a sensitive measure of asthma, but not of rhinitis or urticaria (Newman Taylor et al., 1981).

Work with skin testing has extended beyond the routine measurement of size and character of skin reaction. As mentioned above, skin biopsies are routinely used to examine for the presence of granulomas after a Kveim test; studies are underway to characterize the earlier stages of the inflammation (Mishra et al., 1986). The studies have included examination of inflammatory cell population and mediators in the biopsies of skin lesions during various phases of immediate skin reaction and have led to a better understanding of early pathologic response. During the late-phase reaction, skin biopsies can show neutrophil and lymphocyte influx (Felarca and Lowell, 1971).

A novel method is the injection of antigen into bullae in the skin. This particular challenge allows one to measure the influx of mediators, including histamine, into the site of a skin reaction (Warner et al., 1986).

In summary, skin testing has several advantages, including low cost, wide applicability, ready acceptance by patients, and relative safety. Its major drawbacks include difficulty in assessing observations regarding skin reactions and in correlating reactions with symptoms in other organs and identifying proper antigens for testing.

Nasal Challenge

The upper airways, especially the nose, are a major target of toxic damage. Rhinitis is a common complaint after exposure to toxicants; but research into rhinitis has been limited, because it is not associated with substantial morbidity and its relationship with lower respiratory symptoms is not clear. Nasal challenges do provide information that might not be obtainable with any other method and thus should always be considered when examining new ways of studying toxicants are being examined. Nasal challenges date at least to 1873, when Blakley placed grass pollen in the noses of allergic patients and induced the signs and symptoms of allergic rhinitis (Naclerio et al., 1983).

The method of intranasal challenge varies. Again, identification of the correct antigen is difficult. One method is to study patients with known intradermal reactions. The specific known antigens are then delivered by nebulizer (Naclerio et al., 1983) or by direct application of an extract (Naclerio et al., 1985). Methods of assessing inflammatory reactions after nasal challenge also vary. They include measurement of airway resistance in the challenged nostril (McLean et al., 1976), measurement of mucus production (Malm et al., 1981), subjective assessment (Connell, 1979; Naclerio et al., 1985), and objective assessment of hyperemia and stenosis (Naclerio et al., 1983).

Nasal challenge is fairly safe. The usual symptom is rhinitis, and an occasional patient develops wheezing. Unlike skin tests, it has not been used in large populations. But there is little to suggest that it could not be performed in a similar manner, with the patient observed for some period after challenge. The cost would depend on the extent of assessment. For example, if measurement of mediators in nasal washes were the goal, the assays could become expensive and thus impractical for screening large populations. Observation for the presence of edema would be simple, although difficult to measure. Nasal airway resistance can be measured by anterior rhinomanometry, which is relatively simple and inexpensive (McLean et al., 1976; Naclerio et al., 1983).

In summary, nasal challenge has distinct advantages over skin testing, because it uses a mucosal surface. Direct observation can be used to assess inflammatory response, so it might be appropriate for screening large populations. In addition, when more objective data are required, nasal airway conductance is easily measured.

Intrabronchial Challenge

The best method for assessing airway response to an antigen would be direct observation. The antigen is chosen on the basis of intradermal response. The antigen dose, described in protein nitrogen units (1 unit is the amount that causes a 4 x 4-mm wheal after intradermal injection), is determined. Intrabronchial challenge is then begun at one-hundredth of that unit. Intrabronchial challenge is usually complemented by BAL in the contralateral lung and in the edematous bronchus after challenge.

Bronchoscopy is performed in the usual manner. Subjects are premedicated with atropine, metaproterenol (a beta agonist), and topical Xylocaine. The bronchoscope is advanced to a subsegmental bronchus, the initial dose of antigen is injected through the bronchoscope, and the bronchus is observed for 3 minutes. If there is no change in the bronchus, the dose is increased. A recordable response consists of blanching, edema, or narrowing of the airways.

The major advantage of intrabronchial challenge is its specificity for identifying an inflammatory response in the bronchus. Visualization lasts for only 3-5 minutes, so it would detect only an immediate response. However, repeat bronchoscopy has been done 2-3 days after bronchial challenge to assess persistent changes, and persistent abnormalities in the cell population have also been observed in BAL fluid (Metzger et al., 1987).

Patients challenged to date have been challenged only with antigen to which they have a good skin response. Patients were usually far more sensitive to intrabronchial than to intradermal exposure. Of 11 patients studied by Metzger et al. (1987), nine responded to less than one-twentieth of the intradermal dose.

Intrabronchial challenge presents many problems, mostly because it is relatively new. Although the bronchial changes are visually dramatic, there is little objective measure of response. Because of problems with parallax from a flexible fiberoptic bronchoscope, it is difficult to determine size in the bronchus without a reference object at the same plane as the area one wishes to measure. That is commonly provided by touching the area with an open biopsy forceps or attempting to pass a bronchoscope or a bronchoscopy brush through a narrowed bronchus (Zavala, 1978). Obviously, touching the walls of the bronchus that one wishes to evaluate for edema can lead to local trauma and incorrect interpretation of edema. In addition, accurate estimation of airway narrowing might not be possible when a bronchus has responded to an antigen.

Intrabronchial challenge poses a substantial risk in some persons. The risk associated with bronchoscopy in asthmatic subjects is dealt with in the section on BAL; the risk associated with intrabronchial challenge conceivably is even higher. In the studies reported so far, patients have been carefully selected, many precautions have been observed, and established guidelines have been followed (NHLBI, 1985). Patients have been observed closely for evidence of bronchospasm. In one study (Metzger et al., 1987), three of 11 asthmatic subjects developed wheezing; two were treated with local epinephrine, and the other with aerosol therapy. Pulmonary function of all asthmatics returned to normal within 15 minutes of the procedure.

A final problem in intrabronchial challenge is cost. With the current system, including close observation, studies are expensive and require highly trained medical and technical assistants.

In conclusion, the utility of intrabronchial challenge as a screening tool for identifying patients sensitive to pulmonary toxicants seems limited. In studies to date, only patients who were highly responsive to skin tests responded to intrabronchial challenge. In most of the reported studies, patients were challenged with an antigen clearly associated with pulmonary symptoms. Although the research data obtained after intrabronchial challenge are considerable, their application to a large group of subjects remains questionable.

In Vitro Challenge

Proliferation of lymphocytes exposed to antigen in vitro is an indication of sensitization. In general, lymphocyte proliferation requires the participation of accessory cells and products of Type I or Type II histocompatibility antigens expressed on accessory cell surfaces. Accessory cells are usually macrophages, but dendritic cells, B cells, and perhaps other cells (such as fibroblasts) can act as accessory cells. The exact relationships between lymphocytes and accessory cells in the lung remain to be defined.

The pulmonary lymphocytes obtained with BAL are functionally competent—they can proliferate and produce lymphokines when exposed to antigens to which they are sensitized (Schuyler et al., 1978; Moore et al., 1980; Pinkston et al., 1983). The exact population of lymphocytes resulting from proliferation and the level of lymphokine secretion are not known. Proliferation of antigen-induced and mitogen-induced BAL lymphocytes is lower than proliferation of peripheral blood lymphocytes.

Increases in the percentage and number of BAL lymphocytes are characteristic of granulomatous lung diseases, such as hypersensitivity pneumonitis, sarcoidosis, berylliosis, and tuberculosis (Reynolds et al., 1977; Rossman et al., 1978; Godard et al., 1981; Epstein et al., 1982).

Recent reports indicate that pulmonary lymphocytes from patients with sarcoidosis spontaneously secrete interleukin-2 (Pinkston et al., 1983), which provides a signal for responsive lymphocytes to proliferate. There is evidence that Il-2 secretion by pulmonary lymphocytes from patients with sarcoidosis is secondary to an altered milieu in the lung, rather than being a reflection of changes of the constitutive properties of T lymphocytes (Muller-Quernheim et al., 1986).

Pulmonary lymphocytes from patients with hypersensitivity pneumonitis are sensitized: they proliferate and produce lymphokines on exposure to the appropriate antigen (Schuyler et al., 1978; Moore et al., 1980). There is evidence that cells that suppress lymphocyte proliferation are present in asymptomatic exposed persons, but not in symptomatic exposed persons (Amrein et al., 1970). Therefore, lack of suppressor cells in some subjects could be associated with development of symptoms of hypersensitivity pneumonitis after the same amount of systemic exposure that does not cause clinical symptoms in subjects with suppressor cells.

In general, lymphocytes from patients with berylliosis, but not from control populations, proliferate when exposed to beryllium salts. The results with subjects exposed to beryllium but without apparent disease are controversial (Hanifin et al., 1970; Deodhar et al., 1973; Epstein et al., 1982; Williams and Williams, 1982, 1983; Rom et al., 1983; Bargon et al., 1986). The relationship of lymphocyte proliferation and berylliosis is complex. Beryllium salts have multiple effects on lymphocytes in culture: at high concentrations, beryllium is toxic to lymphocytes and decreases proliferation; at low concentrations, it increases mitogen- and antigen-induced proliferation (Williams and Williams, 1982). Lymphocyte proliferation in peripheral blood has been found to correlate with beryllium exposure in a beryllium plant (Rom et al., 1983) and thus might be a good marker of a population's exposure to beryllium. Although proliferation of lymphocytes from peripheral blood has been studied most extensively, there is preliminary evidence that bronchoalveolar lymphocytes from a patient with berylliosis also proliferate when exposed to beryllium (Epstein et al., 1982).

In summary, lymphocyte proliferation seems to be an index of exposure to environmental agents and in some instances a marker of disease. The relationship of lymphocyte proliferation and pathogenesis in humans is unknown.

6

Markers of Cellular and Biochemical Response

In recent years, much has been learned concerning the cellular and biochemical mechanisms of lung response to both chemical insult and disease. This chapter examines the rapidly developing field of cellular interactions and biochemical mechanisms of respiratory response. It focuses particularly on the analysis of respiratory tract fluids.

SOURCES OF RESPIRATORY TRACT MARKERS

Although this report deals with several possible sources of biologic markers, the introduction of sampling techniques peculiar to the lung and upper respiratory tract has improved understanding of the lung in normal and diseased states. Those techniques are relatively new and still entail some problems in their application for studying biologic responses in large groups of people. This section reviews three techniques for sampling the respiratory tract.

Bronchoalveolar Lavage in Humans

The technique of bronchial washing is not new; Reynolds and Newball in 1974 described a method of bronchoalveolar lavage through a flexible fiberoptic bron-

choscope (Reynolds and Newball, 1974). Their general method has since become widely accepted and is used in many diseases. Bronchoalveolar lavage is the subject of two recent reviews (Daniele et al., 1985; Reynolds, 1987); the following discussion is limited to questions regarding its application to patients or populations exposed to pulmonary toxicants.

Bronchoalveolar lavage (BAL) is usually performed on subjects who are awake. Bronchoscopy with a flexible fiberoptic bronchoscope requires only minimal premedication, usually atropine, and mild sedation. During the procedure, topical anesthesia is provided with Xylocaine (lidocaine). Xylocaine alters the function of alveolar macrophages (Hoidal et al., 1979), but the dose or amount of Xylocaine in the final BAL fluid is usually far below that associated with any effect on alveolar macrophage function (Reynolds, 1987). The bronchoscope is usually passed as far as possible in the right middle lobe or left upper lobe of the lung. Normal saline solution is introduced and aspirated; aspirated fluid is collected and analyzed.

One of the major difficulties in interpreting the literature on BAL findings has been the variety of techniques used for lavage. High-pressure suction (pressure, over 40 cm H_2O) usually leads to airway collapse and poor sampling of the alveoli and therefore to preferential sampling of the bronchi. Several groups have noted that changes in the volume used for lavage result in chemical and physiologic differences in the sample obtained. The first 20-60 ml of instilled fluid usually yields a sample of only the proximal airways, and not the alveoli. Several groups discard the fluid retrieved after the first 20 ml is instilled. When the total volume of instilled lavage fluid was 240 ml, the relative proportions of neutrophils and lymphocytes decreased from the first 120 ml to the second 120 ml in normal subjects. In patients with interstitial lung disease and presumably inflammatory cells in the alveoli, the percentages of lymphocytes and neutrophils increased in the second 120 ml (Dohn and Baughman, 1985). Different portions of the lung might yield different proportions of cells, despite the appearance of a homogeneous disease state, as in sarcoidosis (Cantin et al., 1983) and idiopathic pulmonary fibrosis (Garcia et al., 1986).

Another major problem in BAL is that the source of cells retrieved is unknown. Early studies showed a correlation between the extent of inflammation detected with lavage and later biopsy specimens (Crystal et al., 1981; Paradis et al., 1986). However, results of functional studies have suggested that cells retrieved by BAL differ from those found in the interstitium (Weissler et al., 1986).

Despite the potential wide variability in performing lavage, consensus on how to perform the technique seems to be growing. A questionnaire on BAL technique was completed and returned by 62 centers throughout the world (Klech et al., 1986). Table 6-1 shows good agreement. The variability among centers could well decrease with time.

The amount of fluid withdrawn in BAL is not standard. One usually retrieves 40-80% of the instilled fluid. The aspirated fluid is a mixture of the instilled fluid and lung fluid. There is no satisfactory way to calculate the extent of dilution of instilled fluid with lung fluid.

Markers based on BAL have included endogenous and exogenous markers. Of the endogenous markers, albumin and total protein have been most commonly used. The results of BAL are corrected to milligrams of protein or albumin. In inflammatory states, there is an increase in protein transfer across the alveolar-capillary barrier and therefore an increase in the amount of albumin in the lavage fluid. The increase has been detected in sarcoidosis (Baughman et al., 1983), asthma (Crimi et al., 1983), and oxygen toxicity (Davis et al., 1983). The use of albumin is therefore unsatisfactory in studying disease states not associated with inflammation. Another endogenous marker is urea (Rennard et al., 1986). Urea readily crosses the alveolar-capillary membrane and therefore is in the same concentration in the lung fluid as in the peripheral blood. Although measurement of urea in aspirated BAL fluid would yield some idea of the amount of lung fluid retrieved, there are again problems. Urea passes rapidly from blood into the alveoli, so the longer the lavage tube is in place, the more urea will go into the alveolar space (Sietsema et al., 1986;

TABLE 6-1 Results of Survey on BAL Technique in 62 Centers in 19 Countries

Technique	Proportion of Centers Using Technique, %
Flexible bronchoscopy with only local anesthesia	93
Lavage of either right middle lobe or left upper lobe	98
Use of 100-300 ml of lavage fluid	92
Collection of fluid by pump (low pressure)	77
Collection of fluid in plastic vessels or silicone-coated glass vessels	100
Use of total cell counts	91
Use of differential cell counts	100

Marcy et al., 1987). Apparently, if care is taken to have a short enough dwell time, the concentration of urea in the BAL fluid can be used to estimate lung fluid (Rennard et al., 1986).

One of the exogenous markers studied has been methylene blue (Baughman et al., 1983), a dye that appears to move across the alveolar-capillary membrane relatively slowly. Inulin, a well-studied substance shown to cross poorly (half-time, 72 hours) from the alveoli into tissue (Normand et al., 1971) and methylene blue were comparable exogenous markers for calculating the dilution of instilled fluid by lung fluid. Methylene blue has the advantage of being easily measured, but it interferes with other proteins in the BAL fluid; for example, it is an oxidant and can interfere with measurements of $alpha_1$-antitrypsin activity (Richter et al., 1986).

Cells from BAL have usually been studied with the use of cytocentrifuge-prepared slides or Millipore filters. The cytocentrifuge permits rapid, simple preparation of cells for staining (Baughman et al., 1986a). Results of studies with Millipore filter preparation of cells have suggested that the number of lymphocytes in the BAL fluid is underestimated with the cytocentrifuge technique (Saltini et al., 1984). Several centers are comparing the two techniques in a wide range of patients.

A more costly method of characterizing cells uses a flow cytometer. It allows fluorescent tagging with specific monoclonal antibodies to antigens on the cell surface and has been applied most commonly to characterize T-lymphocyte subpopulations (Hoffman et al., 1980). For example, lymphocytes retrieved by BAL from patients with active sarcoidosis have a high number of T-helper/inducer (Th/i) cells (Crystal et al., 1981); lymphocytes retrieved from patients with hypersensitivity pneumonitis have a marked increase in the number of T-suppressor/cytotoxic (Ts/c) cells (Leatherman et al., 1984). Thus, the ratio of Th/i to Ts/c cells readily distinguishes between active sarcoidosis and hypersensitivity pneumonitis.

The difficulty with fluorescent tagging is the large amount of autofluorescence of alveolar macrophages. The overlap can be minimized with a flow cytometer that has forward and side scatter; macrophages that are larger and more granular than lymphocytes can be removed, and the fluorescence of lymphocytes alone can be studied (Ginns et al., 1982). An alternative is to count cells directly, being assured that they are lymphocytes and seeing how many of them stain. However, hand counting will involve far fewer than the 10,000 cells routinely counted by a flow cytometer.

The risk associated with BAL has been well studied. In general, the risk is not much greater than that associated with routine bronchoscopy (Cole et al., 1980; Strumpf et al., 1981). After the procedure, patients usually have a troublesome cough that resolves quickly, and 10-50% of subjects have a one-time fever that is not associated with infection. Hypoxemia has been noted. Its extent is similar to that observed after routine bronchoscopy, and it is generally measured as a decrease in pO_2 of 10-15 torr that is correctable with oxygen treatment. The hypoxemia can persist for several hours after the procedure. In patients without baseline hypoxia, the decrease in pO_2 after routine bronchoscopy is not clinically significant.

Bronchoscopy can induce bronchospasm (Sackner et al., 1972). Although bronchoscopy is often successful in asthmatic patients who require it, the safety of BAL in asthmatics was the subject of a conference held in 1984 (NHLBI, 1985), which yielded a series of recommendations. Several groups have reported the safety of BAL in asthmatics (Crimi et al., 1983; Joseph et al., 1983; Diaz et al., 1984). Rankin et al. (1984) were not able to demonstrate a change in pulmonary function after BAL in asthmatic subjects. Kirby et al. (1987) performed methacholine challenges a week before and a day after BAL in 10 asthmatics and could not demonstrate changes in reactivity. Several groups have reported that BAL produced no specific morbidity in asthmatic subjects. In some of the studies, the patients were specifically challenged with allergen before or during bronchoscopy (Metzger et al., 1985, 1987).

Although those studies were done on small numbers of patients, their results suggest that BAL can be performed safely on asthmatics. However, asthmatic patients should be selected with care and carefully monitored after the procedure.

Most researchers exclude from BAL patients with moderate to severe airway obstruction due to asthma. Patients with an $FEV_1:FVC$ ratio of less than 0.60:1 are also usually excluded (NHLBI, 1985; Metzger et al., 1985). That approach is safer in that it reduces the likelihood that a patient will develop bronchospasm during the procedure. It has been demonstrated that patients with moderate to severe airway obstruction regularly have a poor return of instilled fluid during BAL (Finley et al., 1967; Martin et al., 1985), probably because of airway collapse during aspiration of the fluid, which is more likely in patients with severe obstructive airway disease.

BAL has been widely applied to evaluation of interstitial lung diseases. Patients with hypersensitivity pneumonitis have a marked influx of lymphocytes into their BAL fluid (Reynolds et al., 1977; Weinberger et al., 1978). The cells are usually characterized as Ts/c cells, so there is a reduction in the Th/i:Ts/c ratio (Leatherman et al., 1984). The Ts/c lymphocyte concentration is clearly higher than that in the normal population, but might not be very different from that in subjects exposed to the same antigen but not ill.

Leatherman et al. studied pigeon breeders and found that those with hypersensitivity pneumonitis had increased lymphocytes in their BAL. They also found that asymptomatic pigeon breeders had increased lymphocytes in their BAL fluid (Leatherman et al., 1984). In studying patients with farmer's lung, another type of hypersensitivity reaction, Cormier et al. (1987) found an increase in the percentage of lymphocytes with acute disease. However, the increase in lymphocytes was found also in patients who continued to work on their farms but had no further symptoms. The authors concluded that BAL lymphocytosis had no prognostic significance for farmer's lung patients.

BAL is useful in securing alveolar macrophages (AMs) from the lung, and retrieval of those cells can be useful in characterizing what the lung has been exposed to. For example, BAL fluid from workers exposed to asbestos might contain ferruginous bodies. The most striking example of changes in the cells in BAL fluid is seen in patients who have smoked cigarettes (Finch et al., 1982). Cigarette-smoking grossly changes the number and properties of AMs retrieved in BAL fluid. There is usually a 10-fold or greater increase in the concentration of AMs retrieved from heavy smokers, compared with nonsmokers. AMs from smokers contain a large amount of amorphous material, which still appears in AMs from ex-smokers. The surface properties and histochemical staining of the AMs have changed. They are also more biochemically active. For example, AMs from cigarette-smokers often spontaneously release hydrogen peroxide and other oxygen radicals (Hoidal and Niewoehner, 1982; Baughman et al., 1986b). In assessing the BAL fluid of patients exposed to pulmonary toxins, one must bear in mind that the changes in the AMs caused by smoking can mask other changes due to toxicants.

Studies have demonstrated the utility of BAL in assessing patients with asthma. Lavage takes place immediately after challenge or later. The delayed lavage has tended to be 6-8 hours after challenge, to correspond to the late phase of the asthmatic response (De Monchy et al., 1985), or 48-96 hours after challenge (Metzger et al., 1985), to determine the presence or absence of persistent abnormalities in BAL fluid. In asthmatic patients who have the biphasic response to antigen, a difference in the BAL-fluid cellular population between the early and late phases can be demonstrated (De Monchy et al., 1985). Eosinophils seem not to appear in BAL fluid until the late phase of a reaction; BAL fluid from patients without a late-phase reaction does not contain eosinophils. That difference supports the current concept that what causes the early phase of the asthmatic response is the release of histamine from mast cells, whereas the late phase is mediated by inflammatory cells (Booij-Noord et al., 1971).

In summary, BAL is an interesting diagnostic tool that allows the sampling of distal airways in a way achieved by no other method. The sensitivity of BAL for disease is not known. It is clear that BAL is not specific for disease, inasmuch as abnormalities can be seen in the BAL fluid of asymptomatic patients (Leatherman et al., 1984; Cormier et al., 1987). In the hands of properly trained personnel, it is a safe procedure. In high-risk patients, it might be a useful way of revealing early biologic effects or altered structure, but its role in screening for disease could be limited, because it can be applied only to select populations.

Bronchoalveolar Lavage in Animals

Analysis of BAL fluid for biologic markers of pulmonary conditions has been useful in animal toxicity studies (Beck at al., 1982; Henderson, 1984, 1988a,b; Henderson et al., 1985a). In large laboratory animals, such as dogs and nonhuman primates, in vivo BAL is usually performed in a manner similar to that used in humans. A fiberoptic bronchoscope is wedged into an airway, and the bronchoalveolar space distal to the wedge is lavaged several times (commonly five or six times) with physiologic saline solution (Muggenburg et al., 1972, 1982). Lavage volumes vary, but 10 ml is adequate. Lavage of small laboratory animals can be performed in vivo (Mauderly, 1977) if required, but most lavages of rodents are performed on excised lungs. A syringe inserted into the trachea is used to instill the saline solution. Either the total lung or a known fraction of it is lavaged. Lavage volumes are usually approximately half the total lung capacity of the section of the lung lavaged. The number of lavages depends on the objective of the study. If the objective is to evaluate the cellular portion of the BAL fluid, numerous lavages, sometimes accompanied by gentle massage of the lung, might be used to retrieve the maximal number of cells; Fels and Cohn (1986) have reported that the most functionally active cells are retrieved in the later lavages. If the objective is to evaluate the acellular fraction of the BAL fluid, two to four lavages might be performed to avoid excessive dilution of the biochemical components to be assayed.

Recovery of lavage fluid in total-lung lavage in control animals is approximately 75% for the first lavage and 100% for later lavages (Henderson, 1988b). In segmental lavages in large animals, the recovery might be less than 50% on the first lavage, but approaches 100% on later lavages. Data from BAL-fluid analyses can be reported in terms of the total amount of fluid retrieved per lung or per gram of lung (if the experimental procedure has not affected lung weight) or the concentration of the constituent of interest in the fluid.

BAL-fluid analysis has been used for a variety of research objectives. The most common use has been to rank various airborne materials for potential pulmonary toxicity by determining the inflammatory lung response that follows administration of increasing amounts of them. A second important use has been to follow the progress of a pulmonary condition in an animal without having to kill the animal. BAL-fluid analysis has also been used to elucidate pathogenic mechanisms in experimentally induced lung disease. Examples of each kind of application are described below.

BAL-fluid analysis has been used to rank inhaled or intratracheally instilled mineral dusts for toxicity (Moores et al., 1980; Morgan et al., 1980; Beck et al., 1981, 1982, 1987; Begin et al., 1983; Henderson et al., 1985b) and similarly to rank metallic compounds for toxicity (Henderson et al., 1979a,b; Benson et al., 1986). The toxicity of an administered material was evaluated according to the degree of inflammation and cell injury as measured by BAL-fluid content of neutrophils (marker of influx of inflammatory cells), serum proteins (marker of increased permeability of the alveolar-capillary barrier), lactate dehydrogenase (marker of cytotoxicity), and lysosomal enzymes (usually either beta-glucuronidase or N-acetyl-beta-glucosaminidase, markers of activation or lysis of phagocytic cells). The degree of increase in those markers in BAL fluid was shown to distinguish between the pulmonary response to more fibrogenic materials (quartz and asbestos) and to less fibrogenic materials (Al_2O_3, Fe_2O_3, latex beads, and fly ash); between the pulmonary response to the highly toxic $CdCl_2$ and to the less toxic $CrCl_3$; and between several nickel compounds in acute pulmonary toxicity. The studies were conducted in sheep, rats, or hamsters; results were similar in all species. In several of the studies, comparisons were made between the histologic evaluation of the effects of the materials and the effects as evaluated by BAL-fluid analysis. The histologic evaluations confirmed the pulmonary conditions. BAL-fluid analysis is a valid means of detecting an inflammatory response in the lung.

Markers in BAL fluid have also been used to measure pulmonary responses to inhaled O_3 (Guth et al., 1986) and NO_2 (DeNicola et al., 1981). The most sensitive biologic markers of the inflammation induced by those gases were increased numbers of neutrophils and, in the acellular fraction, increased protein. Other potential markers of inflammatory response that could be measured in BAL fluid include additional factors released by phagocytic and epithelial cells, such as growth factors, arachidonate metabolites, and interleukin I, which are beginning to be used in toxicology (Seltzer et al., 1986; Henderson et al., 1985a; Koren et al., 1989).

In larger animals, such as dogs and nonhuman primates, BAL-fluid analysis offers a means of following the course of a pulmonary condition sequentially in the same animal. By inserting the bronchoscope in different airways, one can perform BAL several times in the same animal without lavaging the same area. Or one can instill a test material into one area and a vehicle into another area and use a given animal as its own control in determining the effect of the test material. Both applications have been used by Bice et al. (1980a) and reviewed by Bice (1985). The investigators instilled sheep red blood cells into the left lung of a dog and followed the course of appearance of IgM- and IgG-forming cells. The right lung of the same dog received saline solution and served as a control. Significantly more antibody-forming cells were found in BAL fluid from the immunized lung than from the control lung. One could even do a whole dose-response study in the same animal by instilling different amounts of the test material into different areas of the lung. In one study (Bice and Muggenburg, 1986), various numbers of sheep red blood cells were instilled into different areas of a dog's lung to determine the dose-response characteristics of the immune response.

Bice et al. (1982) used BAL-fluid analysis to elucidate the mechanism of recruitment of immune cells to the lung. Two lung lobes of a dog were immunized with antigenically different particles (sheep and rabbit red blood cells) to determine whether immune cells in the blood are recruited to the lung in an antigen-specific manner. Analysis of BAL fluid indicated that equal numbers of anti-sheep-red-blood-cell antibody-forming cells were present in both immunized lobes. The authors concluded that cell recruitment was not antigen-specific, but related to nonspecific changes in the lobes induced by antigen exposure. Another example of the usefulness of biologic markers in BAL fluid in the elucidation of mechanisms of disease is the work of Holtzman et al. (1983), who found that an increase in airway responsiveness in dogs was related to an influx of neutrophils and increases in prostaglandins E2 and F2a in BAL fluid.

Thus, the markers of biologic events that can be found in BAL fluid are useful in toxicology. The use of BAL-fluid markers has several advantages. BAL-fluid analysis results in a rapid, quantitative measure of pulmonary response that is not obtained with routine histologic evaluations. BAL-fluid analysis allows detection of early biologic events. Investigators have reported detection of inflammation through BAL-fluid analysis before radiographic detection was possible (Fahey et al., 1982). In large animals, the procedure can be used sequentially to follow the course of a biologic event in a given animal.

The present limitation of the method is the lack of specificity of the markers for the site of inflammation or injury or for the lung disease. Only a general inflammatory response can be detected. Continued research is needed to develop site-specific markers of respiratory tract injury and validation of profiles of BAL-fluid changes that are indicative of the presence of or progression toward a specific lung disease. Especially useful would be markers to indicate the early stages of a progressive condition that leads to disease, such as fibrosis, emphysema, or cancer. In toxicology, such markers would allow earlier detection of late-occurring events; if they were applicable to humans, they would allow therapeutic intervention at an early stage in a disease process. Research to elucidate the mechanisms by which respiratory diseases develop should aid in obtaining the information required to select the correct markers.

Bronchial Lavage

Bronchial lavage, a variation of BAL, has emerged in the last few years. The need for such a procedure became obvious as people began to note the differences between small- and large-volume lavage (Dohn and Baughman, 1985). The cells retrieved in the first portions of BAL fluid are from the larger airways. Although those cells might reflect contamination in subjects with alveolar disease, the first portions might be of most interest in connection with patients in whom the disease is of the large airways.

In one method of bronchial lavage, a catheter is passed through a flexible fiberoptic bronchoscope into a main bronchus with light anesthesia. Attached to the outside of the catheter are two balloons several centimeters apart. The balloons are blown up, occluding the airway and sealing the section of bronchus between the two balloons. Fluid is then introduced into and withdrawn from the lumen between the balloons. Eschenbacher and Gravelyn (1987) used the method to expose the bronchial wall to hypo-osmolar challenge, lavage the area, and examine the fluid for biochemical factors.

Nasal Lavage

The nose is the primary portal of entry of inspired air, and one of its major roles is to protect the lower respiratory tract from inhaled pollutants. For example, 100% of SO_2 drawn into the nose, 20-80% of O_3, and 73% of NO_2 are trapped there under normal conditions (Vaughan et al., 1969). Therefore, if nasal clearance is impaired, a larger amount of pollutants could reach the lower lung. That is reason enough to study the effects of pollutants on the nasopharyngeal region; another reason is that the nasal passages contain many of the same cell types as the trachea and bronchi, but are more convenient and more accessible for studying in vivo effects of airborne toxicants (Proctor, 1982; Cole and Stanley, 1983; Koenig and Pierson, 1984).

Secretions from the nasal area have been analyzed for various proteins and cell types (Remington et al., 1964; Rossen et al., 1966; Lorin et al., 1972; Mygind et al., 1975). In clinical trials, one cannot ensure that rhinitis will be produced in experimental subjects; even if a pollutant is noxious enough to produce a heavy secretion, an unexposed control group will provide little secretion for comparison. An alternative is to collect specimens with nasal lavage. Nasal lavage is simple to perform, noninvasive, and nontraumatic. In an adaptation of the technique reported by Powell et al. (1977), the subject is instructed to sit upright with head tilted back and to establish palatal pressure. A needleless syringe is used to instill into each nostril 5 ml of sterile phosphate-buffered saline solution (Brain and Frank, 1973). The saline solution is held in the nasal passages for 10 seconds and then forcibly expelled and collected. Of the 10 ml instilled, about 7 ml is routinely recovered.

Sloughed squamous, columnar, and (less often) ciliated epithelial cells are routinely found in nasal lavage. Leukocytes are normally present, and their numbers increase in some disease states. An increase in nasal eosinophils has been used as a clinical verification of an allergic reaction (Malmberg and Holopainen, 1979; R. E. Miller et al., 1982); and nasal basophils have been shown to increase in allergic persons 10-20 minutes after antigen challenge (Bascom et al., 1988). Neutrophils increase by a factor of 10-100 during an upper respiratory tract viral infection (Farr et al., 1984; Henderson et al., 1987). A significant increase in nasal-lavage neutrophil numbers has also been shown to occur in response to acute exposure to ozone at 0.5 ppm; ozone is an oxidant air pollutant known to induce an inflammatory response in the lungs of animals (Graham et al., 1988).

When looking at changes in the nasal-lavage cell population, one must consider the effect of earlier unidentified environmental exposures. Of 200 volunteers, 50% had fewer than 10^4 polymorphonuclear neutrophils (PMNs) per milliliter of nasal lavage, and 10% had over 10^5 PMNs per milliliter (Graham et al., 1988) The remaining lavages were evenly scattered between those extremes. Responses to a questionnaire on life style and exposure suggest that increased numbers of PMNs might be associated with recent colds, with exposure on the previous evening to heavy cigarette smoke or chemicals found in Chlorox and paint stripper, with gasoline fumes, or with recent swimming in lakes or ponds. Such environmental exposures could account for the variability in cell counts seen by Farr et al. (1984) when five samples were taken from the same person over a 2-week period. Potential effects of uncontrolled environmental exposures must be taken into account in the design of a study. Pre-experiment and post-experiment samples taken on the same day and instruction of subjects can reduce the confounding effects.

Markers in nasal-lavage fluid that have been studied include increase in total protein, associated with cell damage and permeability change (Lorin et al., 1972; Marom et al., 1984); increases in concentrations of albumin and immunoglobulin G, associated with increased vascular permeability (Butler et al., 1970; Rossen et al., 1971; Brandtzaeg, 1984); histamine and prostaglandin D_2, released from mast cells in response to an allergic reaction (Naclerio et al., 1983; Eggleston et al., 1984); increases in concentrations of sulfidopeptide leukotrienes and kininogens, after an allergic response (Creticos et al., 1984; Baumgarten et al., 1985; Togias et al., 1986); and increases in concentrations of immunoglobulin E in hay fever (Miadonna et al., 1983; Small et al., 1985).

Substances measured in nasal-lavage fluid have an unknown dilution factor, as is the case in BAL fluid. The absolute concentration cannot be determined, and interpretations are of limited value. Pre-experiment concentrations of mediators can vary widely between individuals and generate a "noisy" baseline. In studying allergic responses, Togias et al. (1986) uses four or five lavages before an experimental exposure and analyzes the first and last of them. That yields information on the pre-experiment concentrations and provides a more stable baseline. Analysis of nasal-lavage fluids for specific mediators might thus not be practical for screening large populations.

In spite of those concerns, much information can be obtained from nasal lavage. The procedure allows measurement of an effect of a pollutant on a mucosal surface, requires no anesthetic, and does not itself induce the release of mediators (Baumgarten et al., 1985) or an inflammatory response (Graham et al., 1988). Multiple samples from the same person are possible, as well as samples from subjects who are at risk, such as asthmatics. No special equipment is required, so this is an attractive and inexpensive approach for epidemiologic or occupational studies. Furthermore, nasal lavage can be useful in determining which air pollutants might induce an inflammatory response in the human respiratory tract. Increases in neutrophil, eosinophil, or basophil concentrations—which are easily measured and have been associated with health effects—could be the most useful markers in the nasal lavages.

More environmental and occupational studies analyzing nasal lavages for both cells and mediators in both normal and asthmatic persons are needed for a full appreciation of the value of this approach in screening. Studies of such different pollutants as sidestream tobacco smoke, cotton dust, and SO_2 could be useful in that regard. Nasal lavages might be useful in studying the effects of indoor air pollution, inasmuch as nasal irritation is one of the most common complaints associated with that pollution. Studies comparing cell and mediator changes in the nasal-lavage fluid with those found in BAL fluid from pollutant-exposed people are needed to determine whether the nasopharyngeal region can be used as a diagnostic mirror of the lower respiratory tract—i.e., whether nasal symptoms can herald lower respiratory disease or provide important clues to coexisting chest disease.

POTENTIAL MARKERS IN RESPIRATORY TRACT FLUIDS

Both the cellular and acellular contents of nasal, bronchial, and bronchoalveolar lavages can provide markers of response to environmental exposures, as shown in Table 6-2. In the following sections, the cellular and the acellular supernatant fractions of these fluids and their potential for use as biologic markers are discussed in more detail. Their use as markers is summarized in Table 6-3.

Cellular Content

Macrophages

The predominant cell in BAL fluid from normal subjects is the macrophage. In some species (Henderson, 1988a), lymphocytes can be present in small numbers. Neutrophils, eosinophils, and mast cells might also be present as a result of an inflammatory response.

Human alveolar macrophages (AMs) are composed of several populations that can be distinguished by density. In general, denser AMs are less mature and resemble blood monocytes more than less dense AMs. The denser cells are more potent producers of a soluble factor that inhibits fibroblast proliferation (Elias et al., 1985a) and of interleukin-1 (Il-1) (Elias et al., 1985b), and they are more efficient accessory cells for antigen-induced proliferation (Ferro et al., 1987). AM abnormalities in sarcoidosis might represent differences in the relative proportions of AM subpopulations, rather than intrinsic differences in the same AM subpopulations.

Hance and colleagues (1985) have found that AMs from patients with sarcoidosis express antigens that are present on blood monocytes, but AMs from normal subjects do not. Other attributes of sarcoid AMs, such as accessory cell function (Venet et al., 1985) and spontaneous release of interferon-gamma and growth factors (Bitterman et al., 1983), are compatible with the influx of less mature AMs in sarcoidosis.

• *Accessory Cell Function.* Macrophages are important in the regulation of the immune response, acting as both promoters and suppressors of events that result in immunization or inflammation (Unanue et al., 1984). In general, AMs that can be obtained from normal humans with lavage contain subpopulations that can increase or suppress lymphocyte proliferation. The result depends on culture conditions—the presence of other accessory cells and the amount of antigen or mitogen present (Liu et al., 1984; Toews et al., 1984; Ettensohn et al., 1986). Alterations in accessory cell function might affect pathogenesis. In fact, as mentioned above, some studies have shown that AMs from patients with sarcoidosis have accessory cell function more efficient than that in AMs from normal subjects (Lem et al., 1985; Venet et al., 1985). AMs from asthmatics are less able than AMs from normal subjects to suppress mitogen-induced lymphocyte proliferation (Aubas et al., 1984). Changes in accessory cell function are gaining acceptance and should be used as markers of environmental exposure and perhaps as markers of susceptibility to disease.

• *Interleukin-1.* AMs can secrete Il-1, but apparently to a smaller degree than can peripheral blood monocytes (Koretzky et al., 1983; Wewers et al., 1984). The uncertainty regarding AM Il-1 production and its control mechanisms probably derives from the multiplicity of AM products, some of which antagonize Il-1 production (Monick et al., 1987). Il-1 is an important mediator of inflammation (Dinarello, 1984) and acts as a differentiation signal for several subsets of lymphocytes. In particular, Il-1 promotes differentiation

TABLE 6-2 Techniques for Detecting Markers of Inflammatory and Immune Response

Technique	Advantages	Disadvantages
Bronchoalveolar lavage	Relatively safe; material obtained from lung (compartmentalization); large quantities of cells and fluid available; repeatable; large area of lung sampled	Variable concentration of return fluid constituents; expensive; obtaining of normal controls in large numbers difficult; sample can be interstitial, alveolar, or bronchial; invasive
Nasal lavage	Suitable for large studies; safe; repeatable; establishment of reproducible baseline values possible; inexpensive; affords both cellular and fluid analysis; quick return to baseline; no anesthesia needed	Yields few cells; might not reflect lower respiratory tract response; response short-lived; timing of sampling after exposure critical; requires patient cooperation and understanding
Blood and urine collection	Inexpensive; safe; repeatable; large sample available; widespread patient acceptance	Often does not reflect respiratory tract response; risk of exposure of investigator
Measurement of cellular influx	Sensitive for inflammatory and immune responses; easily measured by established methods; presence of increased numbers of some leukocyte types is associated with specific pulmonary responses	Leukocyte influx a nonspecific response and might be transient
Phenotyping	Specific; reproducible	Expensive; pathophysiologic relevance of specific marker must be established; might require large number of cells
Measurement of arachidonic acid metabolites, cytokines, and enzymes	Probably important mediators of respiratory tract injury easily measured in respiratory secretions; relatively inexpensive; sensitive	Expensive; difficult with current methods; importance currently unknown; nonspecific; samples sometimes cannot be stored in bulk for later analysis
Examination of mast cell	Sensitive; relatively specific for inflammation	Measurement difficult; expensive
In vitro assay	Measures antigen-specific response; risk low; well established	Cumbersome; expensive; requires extrapolation to whole-body response; relevant antigens unavailable
Skin test	Measures antigen-specific whole-body response; has been used in large population studies	Some associated risk; sensitivity and specificity often poor; relevant antigens unavailable
Intrabronchial test	Measures antigen-specific tissue response	Use still limited; some associated risk; expensive

TABLE 6-3 Biologic Markers of Inflammatory and Immune Response in Humans[a]

Marker	Circumstances			
	Controlled Challenge	Natural Illness	Occupational Study	Environmental Study
White blood cells				
Neutrophil-Lymphocyte				
Influx				
Nasal	Y	Y	P	P
Bronchial	Y	P	N	N
BAL[b]	Y	Y	Y	N
Neutrophil release				
Oxyradicals	P	Y	N	N
Factors	P	Y	N	N
Lymphocyte (BAL)				
Subpopulation changes	Y	Y	Y	N
Functional changes	Y	Y	Y	N
Macrophage (BAL)				
Release of factors	Y	Y	P	N
Release of oxyradicals	Y	P	P	N
Change in cell surface	Y	P	P	N
Mast cells				
Influx	Y	Y	P	N
Release of factors	Y	Y	P	N
Eosinophils				
Influx	Y	Y	Y	Y
Release of factors	P	Y	N	N
Biochemical changes				
Arachidonic acid metabolites				
Urine	N	Y	N	N
Nasal wash	Y	N	N	N
BAL	Y	Y	N	N
Histamine				
Blood	N	Y	N	N
Nasal wash	Y	Y	P	N
BAL	Y	Y	N	N
Enzymes, fluid				
Blood	Y	Y	N	N
Nasal wash	Y	Y	N	N
BAL	Y	Y	P	N
Enzymes, cell-associated				
Nasal wash	P	P	N	N
BAL	N	Y	N	N
Functional changes				
Antigen-toxin challenge				
Skin	Y	Y	Y	Y
Nasal wash	Y	Y	P	N
Aerosol	Y	Y	Y	P
Intrabronchial	Y	Y	N	N
Rast	N	Y	Y	N
In Vitro				
Lymphocyte				
Blood	Y	Y	Y	N
BAL	Y	Y	P	N
Monocyte-Macrophage				
Blood	Y	Y	Y	N
BAL	Y	Y	N	N

[a]Y = yes (well documented); P = preliminary data; N = not done yet.

and maturation of helper T cells (Mizel, 1982), lymphocytes (Rao et al., 1983), and natural killer cells (Dempsey et al., 1982). Il-1 is chemotactic for neutrophils, can alter the adherence properties of endothelial cells by stimulating prostaglandin synthesis (Bevilacqua et al., 1985), and can induce fibroblast proliferation (Schmidt et al., 1982). In animals, perturbation of the lung can cause spontaneous release of Il-1 (Lamontagne et al., 1985). An analogous situation can occur in sarcoidosis, in that Il-1 is spontaneously released in vitro in the absence of additional signals (Hunninghake, 1984). AM Il-1 secretion is influenced by other cytokines, such as interferon-gamma. AMs from normal subjects and persons with lung disease might differ in their response to stimuli that increase Il-1 secretion (Eden and Turino, 1986).

• *Interferon.* AMs from normal subjects can release interferon (-alpha or -gamma) after stimulation by appropriate inducers. It is interesting that cigarette-smoking does not influence inducer-stimulated interferon secretion (Nugent et al., 1985). However, AMs from patients with sarcoidosis spontaneously release interferon without the necessity of inducers (Robinson et al., 1985).

• *Reactive Oxygen Species.* AMs release reactive oxygen intermediates, such as O_2^-, H_2O_2, OH^-, and singlet oxygen. Those products have many proinflammatory effects, including inactivation of protective substances in the lung and cytotoxicity. Changes in the signals required for release of those products or changes in the signals controlling the release of protective substances, such as superoxide dismutase, could have important implications regarding the results of exposure to injurious stimuli. For example, the production of oxygen radicals is increased in AMs from smokers (Hoidal et al., 1981) and from subjects with sarcoidosis (Greening and Lowrie, 1983). Interferon-gamma can increase in vitro oxygen radical production (Fels and Cohn, 1986). AMs from asthmatics differ from those from normal subjects, in that they have increased numbers of IgE receptors and require only exposure to antigen to secrete reactive

oxygen products, whereas AMs from normal persons must first be coated with IgE antibody (Joseph et al., 1983). Release of other AM products, such as platelet activating factor and neutrophil and eosinophil chemotactic factors (possibly leukotriene B_4), exhibit the same dependence on surface IgE antibody (Gosmset et al., 1984). AMs from normal animals do not release reactive oxygen intermediates after stimulation with phorbol esters, unless they are conditioned in vitro with serum; AMs from animals with pulmonary inflammation do not require conditioning to become responsive to phorbol esters (Gerberick et al., 1986). AM sensitivity to stimuli that cause secretion of toxic oxygen intermediates might be a useful marker of lung damage and inflammation.

• *Chemotactic Factors.* AMs release chemotactic factors for neutrophils (Hunninghake et al., 1980a), whose products are important in the pathogenesis of several types of lung injury, such as adult respiratory distress syndrome (Lee et al., 1981; Tate and Repine, 1983) and bronchopulmonary dysplasia. AMs also release factors that stimulate neutrophil adhesiveness and superoxide generation (Tate and Repine, 1983). Cigarette smokers' AMs, unlike nonsmokers' AMs, spontaneously release a neutrophil chemotactic factor in vitro (Hunninghake and Crystal, 1983). Other environmental exposures might similarly change the signals needed to cause release of chemotactic factors.

• *Miscellaneous Factors.* Pulmonary macrophages can release a factor that causes the release of histamine from human basophils and lung mast cells (Schulman et al., 1985). Alterations in the mechanisms that control the release could both cause inflammation and be a result of environmental exposure. AMs release growth factors that influence diverse cell types, such as fibroblasts (Chapman et al., 1984), smooth muscle cells (Martin et al., 1981), endothelial cells (Martin et al., 1981), and Type II cells. Although Il-1 is one of those factors, it is clear that additional substances are released by macrophages (Leslie et al., 1985). The proliferation of many lung cell types that is a common result of inflammatory stimuli could be caused by AM growth factors, and release of the factors could be an early event in lung changes and provide markers of early lung injury.

Neutrophils

Neutrophils are not usually found in large numbers in BAL fluid from normal lungs (Reynolds, 1987; Henderson, 1988a). In lavage fluid from the lungs of smokers (Young and Reynolds, 1984) or from subjects that were exposed to ozone (Seltzer et al., 1986; Koren et al., 1989), an increased percentage and greatly increased absolute numbers of neutrophils are found. As participants in the inflammatory response in lung tissue, neutrophils migrate into the lung from the blood under the influence of one or several chemotactic factors, including complement component C5a and soluble factors secreted by AMs. They might be considered as a secondary line of phagocytic defense for the lungs, which can be recruited into the airspaces in response to exposure to microbial agents or other inhaled materials.

The neutrophil itself contains a host of materials potentially damaging to lung tissue (Henson, 1972; Martin et al., 1981). It has the capacity to release oxygen radicals. In addition, it releases several digestive enzymes, especially neutrophil elastase. The release of neutrophil elastase has been associated with significant lung destruction (Janoff, 1972), as seen in the adult respiratory distress syndrome, where it has been found in lavage fluid (Lee et al., 1981). Neutrophil elastase can cause emphysematous changes in an animal model and has been proposed as a major cause of emphysema. Other less destructive processes have been associated with the release of enzymes from the neutrophil.

Lymphocytes

The lymphocytes in lung fluid usually are T cells, although natural killer cells and B cells have also been found. The importance of B cells and NK cells in the lung is unclear, inasmuch as these cells are susceptible to the suppressive effects of surfactant and other substances in the lung lining fluid. T cells have been divided into two classes on the basis of the monoclonal antibodies directed against antigens found on the surface of the T cells. For example, CD4 cells are in the Th/i cell class, and CD8 cells are in the Ts/c cell class.

Regardless of the cell type responsible for antigen presentation in the lung, antigen must be presented to stimulate T-cell proliferation. It is not clear whether antigen is presented in the lung or in the lymph nodes that drain the lung. Results of investigations in dogs and nonhuman primates suggest that antigen deposited in the lung is translocated to draining lymph nodes (Kaltreider et al., 1977), where proliferation of antigen-specific B cells occurs (Bice et al., 1980b). The antigen-specific B cells then enter the blood and are recruited by nonspecific means to the lungs (Bice et al., 1982). With respect to secondary antigen challenges, results of studies in cynomolgus monkeys suggest that local proliferation of lymphocytes can occur in the lung (Mason et al., 1985). Little is known of primary immune responses in the human lung, but mechanisms responsible for the accumulation of T cells in the lungs of patients with sarcoidosis have been the subject of numerous investigations.

It has recently been shown that T cells in the lower respiratory tract of patients with active sarcoidosis proliferate spontaneously (Pinkston et al., 1983) and release Il-2 (Hunninghake et al., 1983). The major sources of Il-2 are T cells that express human leukocyte antigens-D region (HLA-DR) on their surface (Saltini et al., 1986). Il-2 secretion is probably associated with the pulmonary lymphocyte infiltrate in sarcoidosis. When Il-2 release by lung T cells is suppressed by treatment of patients with corticosteroids, the spontaneous proliferation of T cells ceases, and the numbers of T cells and disease activity subside (Ceuppens et al., 1984; Pinkston et al., 1984; Bauer et al., 1985; Rossi et al., 1985).

A recent investigation determined that Il-2 gene expression by lung T cells was increased in patients with active sarcoid-osis (Muller-Quernheim et al., 1986). The same investigators found that activa-tion of the Il-2 gene in T cells occurred only at the sites of disease and was not a generalized property of T cells through-out the body. Those results suggest a com-partmentalized or local response in the lung, as well as the importance of Il-2 in the pathogenesis of sarcoidosis.

It appears that in lung T-cell spontane-ous proliferation, Il-2 release, Il-2 gene activation, and Il-2 receptor expression are potential markers of lung immune re-sponses. Those markers are inducible and found in patients with active sarcoido-sis, not in normal people. T-cell spon-taneous proliferation and Il-2 secretion are determined with bioassays; because they are cumbersome, the assays are limited in usefulness as markers. As stated ear-lier, Il-2 receptor expression is rela-tively easy to determine and can be a useful marker. Il-2 gene activation can be meas-ured by determining Il-2 mRNA transcripts of lung T cells. The method is technical, but samples can be run with relative ease, once the assay is in place.

Besides Il-2, T cells from patients with sarcoidosis spontaneously secrete macro-phage-migration inhibition factor (Hun-ninghake et al., 1980b), leukocyte in-hibiting factor (Hunninghake et al., 1979a), monocyte chemotactic factor (Hun-ninghake et al., 1980c), and factors that stimulate B lymphocytes to differentiate into immunoglobulin-secreting cells (Hunninghake and Crystal, 1981b; Hun-ninghake et al., 1983). Although those factors are secreted by lung lymphocytes from patients with sarcoidosis, their usefulness as markers is limited, because the assays used to detect their activities are technically difficult.

Increased numbers of lymphocytes in BAL fluid might be associated with nonsympto-matic responses. For example, in a study of 24 dairy farmers with no symptoms of lung disease, 13 had abnormally high percent-ages of lymphocytes in BAL fluid (Cormier et al., 1984). In a followup study of many of the same persons (Cormier et al., 1986), most of the farmers who initially had high numbers of lung lymphocytes still had high numbers 2-3 years later; but none had de-veloped extrinsic allergic alveolitis (EAA). In another study, 5 of 42 nonsmoking normal subjects had lymphocyte propor-tions greater than 20% in BAL fluid (Laviolette, 1985). In a followup examina-tion 47 days later, 4 of the 5 had lymphocyte proportions less than 14%. Those data suggest that the percentage of lymphocytes in BAL fluid fluctuates substantially in normal subjects and that lymphocyte proportions higher than 14% should not necessarily be considered abnormal (Laviolette, 1985).

Characteristically in sarcoidosis and EAA, an increased percentage of BAL-fluid lymphocytes is observed (Reynolds et al., 1977; Hunninghake et al., 1979b; Valenti et al., 1982). As discussed above, lympho-cyte percentage alone cannot be used as an indicator of lung response. However, the enumeration of the phenotypes of lym-phocytes can be more indicative of the im-munologic state of the lung. Lung T cells of patients with active sarcoidosis are characterized by an increased number of Th cells (Hunninghake and Crystal, 1981b; Semenzato et al., 1982). In patients with active disease, the ratio of Th to Ts cells can be as high as 20:1 in both the lung and the hilar lymph nodes (Semenzato et al., 1982; Thomas and Hunninghake, 1987); in normal subjects, the ratio is approximate-ly 1.6:1. However, in patients with inac-tive sarcoidosis, the lung T cells are primarily of the suppressor phenotype, and the ratio of Th to Ts cells is below 1.6:1 (Hunninghake and Crystal, 1981b; Hunninghake et al., 1983).

Th cells are increased in active sar-coidosis, but EAA is characterized by an increase in Ts cells in BAL fluid (Cormier et al., 1980; Moore et al., 1980; Costabel et al., 1984; Leatherman et al., 1984). That increase is also found in persons who have been exposed to the EAA-causing anti-gen but are without symptoms of disease. Only persons who develop EAA have impaired Ts-cell function (Hughes et al., 1984; Keller et al., 1984).

From results of studies to date, it ap-pears that increased lymphocyte numbers in BAL fluid are consistent with the pres-ence of some immune responses in the lung, but do not necessarily indicate that such a response is occurring. Of more usefulness as markers are the numbers of Th and Ts cells. The phenotypes are determined with monoclonal antibodies to lymphocyte surface antigens. It is relatively easy to process multiple samples, because much of the procedure can be automated. However, one must exercise caution in interpreting the meaning of the presence of different phenotypes, inasmuch as the monoclonal antibodies used simply detect surface antigens and do not determine cell function. A case in point is the substantial number of Ts cells in EAA detected with monoclonal antibodies. The result suggests that Ts cells are not involved in the pathogenesis of EAA in symptomatic subjects; however, when Ts-cell activity is measured with a bioassay, a defect in Ts-cell activity becomes apparent in symptomatic subjects.

Eosinophils

Eosinophils are similar to neutrophils. They contain many potentially irritating substances, including eosinophil major basic protein and leukotrienes (Gleich et al., 1979; Weller et al., 1983). Eosinophils appear to have a role in asthma and are often seen in various allergic reactions, including asthma. They are also found as a result of exposure to toxicants, such as the antibiotic bleomycin (Thrall et al., 1979). Chronic eosinophil deposition in the lung might lead to pulmonary fibrosis, although the mechanism is not known (Davis et al., 1984).

Eosinophils have often been used as a marker for specific allergic reactions or toxicant challenge, both in the blood and in respiratory fluids. They are relatively easy to recognize and easy to count. Eosinophil influx into lavage fluid has been seen 6-48 hours after allergen challenge in sensitized asthmatics (De Monchy et al., 1985; Metzger et al., 1987).

Mast Cells

Mast cells are found in the walls of alveoli and airways, usually between the basement membrane and epithelial membrane in normal humans. Mast cells are present in BAL fluid of both normal subjects and those with asthma; the fraction of mast cells in lavage cells is low—less than 0.35% in asthmatics (Tomioka et al., 1984).

Human mast cells, like AMs, can be divided into subpopulations on the basis of density. Recent evidence indicates that there are differences among the subpopulations with regard to histamine content and responsiveness to anti-IgE (Schulman and Anderson, 1985).

Release of mast-cell mediators in vivo presumably occurs after cross-linkage of surface IgE antibodies by specific antigen in a calcium-dependent process. However, human mast cells can be activated to secrete mediators by the complement-derived fragments C3a, C4a, and C5a (Hugli, 1981) and non-IgE stimuli. Some reports note that basal and IgE-induced secretion is greater in mast cells from asthmatics than in mast cells from normal subjects.

Mast-cell mediators are preformed and released after cell stimulation or they can result from biochemical events initiated by stimulation. The preformed mediators are histamine, exoglycosidases (arylsulfatase B, beta-glucuronidase, beta-hexosaminidase B, and beta-galactosidase), proteases (tryptase, carboxypeptidase B, elastase, and cathepsin G), chemotactic factors (eosinophils and neutrophils), heparin, superoxide dismutase, and peroxidase (Metcalfe et al., 1981; Holgate and Kay, 1985). Newly generated mediators after cell initiation are arachidonic acid metabolites, prostaglandins, thromboxane B_2, leukotrienes, and platelet activating factor (MacGlashan et al., 1983).

Histamine increases vascular permeability through venular dilation and creation of endothelial pores, causes bronchoconstriction (especially in asthmatics), and increases prostaglandin generation. It also increases gastric acid and mucus secretion (MacGlashan et al., 1983). Histamine has many effects on immune response whose details are not yet completely understood. In general, it seems to increase early events and inhibit late events in the immune response (Khan and Melmon, 1985). For example, histamine has been

shown to increase lymphocyte proliferation (Thomas et al., 1981), generation
of Ts/c cells, and PMN-stimulated immunoglobulin production (Lima and Rocklin,
1981). Some of the effects of histamine
are mediated by products from other cells,
such as lymphokines. An example of a lymphokine that influences histamine release
is histamine-induced suppressor factor
(HSF), which is produced by T cells after
a series of interactions with macrophages
and their soluble products. HSF suppresses
antigen- and mitogen-induced lymphocyte
proliferation, reduces the release of
other lymphokines, inhibits immunoglobulin synthesis, and suppresses mixed
lymphocyte reactivity (Beer et al., 1984).
Other lymphokines produced by mononuclear
cells activated by histamine also have
lymphocyte chemokinetic properties (i.e.,
they produce lymphocyte chemoattractive
factor, or LCF) or inhibit lymphocyte migration. Mononuclear cells with H_2 receptors produce LCF, and cells with H receptors produce factors that inhibit lymphocyte mobility (Center et al., 1983).

Heparin, another preformed mast-cell
mediator, has many effects, including
anticoagulation. It decreases the capacity of eosinophilic major basic protein
to kill parasites, and it increases protease activity, endothelial cell migration, release of lipoprotein lipase, and
fibronectin binding to collagen (Schwartz
and Austen, 1984).

Prostaglandin D_2 (PGD_2), leukotriene
C_4 (LTC_4), and possibly platelet activating factor are examples of newly synthesized mast-cell factors that are found
in lavage fluid and are secreted by pulmonary mast cells with IgE-dependent mechanisms in vitro (Schwartz and Austen, 1984;
Schleimer et al., 1984). PGD_2 is a potent
bronchoconstrictor and vasodilator that
can potentiate the increase in capillary
permeability produced by other mediators,
such as histamine and leukotrienes
(Holgate and Kay, 1985). LTC_4 is a potent
bronchoconstrictor (Drazen et al., 1980),
increases mucus secretion (Coles et al.,
1983), and increases vascular permeability (Soter et al., 1983). Platelet activating factor causes bronchoconstriction,
increased capillary permeability, and

leukocyte chemotaxis. Tryptase, the major
neutral protease in human mast-cell gran-
ules, generates C3a (Schwartz et al., 1983)
and might be responsible for the inactivation of kininogen after mast-cell activation (Schleimer et al., 1984).

The circulating neutrophil chemotactic
factor (NCF) that appears in plasma after
antigen-induced bronchoconstriction
presumably originates in pulmonary mast
cells, but it might originate in other
cells, such as T lymphocytes and monocytes,
in an IgE-dependent manner (Holgate et
al., 1986).

Acellular Content

Antioxidants

The lung is prone to oxidant injury from
a variety of sources. Such oxidants as
superoxide anion (O_2^-), hydrogen peroxide
(H_2O_2), and hydroxyl radical ($OH^\bullet$) can be
generated in the lower respiratory tract
through the action of PMNs and AMs involved
in local inflammatory reactions or through
the direct effects on lung tissue of inhaling oxygen-rich air or toxic substances
(such as paraquat). In addition to causing
cell and tissue damage, oxidants can inactivate alpha$_1$-antitrypsin, which plays
a major role in defending the lung against
proteolysis from neutrophil elastase and
other proteolytic enzymes released during
inflammation (Carp and Janoff, 1979; Gadek
et al., 1981). The lungs are equipped with
both intracellular and extracellular
protective mechanisms that can combat the
effects of locally generated oxidants.

Alveolar fluid has been shown to contain
a wide spectrum of plasma proteins, some
of which—including transferrin and ceruloplasmin—can function as lower respiratory tract antioxidants. Bell et al.
(1981), for example, have demonstrated
that normal human lavage fluid is rich in
transferrin—the concentration is an
average of 40% higher than that in serum.
They speculated that local synthesis could
be responsible for the enrichment. Human
peripheral blood lymphocytes do synthesize and secrete transferrin, and there
is some evidence that alveolar lymphocytes
come from peripheral blood (Daniele et

al., 1977a). It is not known whether pulmonary alveolar macrophages produce and secrete transferrin. In any event, high concentrations of transferrin at the respiratory surface are advantageous, regardless of its source. Additional studies showed that ceruloplasmin is present in normal BAL fluid. It has been suggested that ceruloplasmin can help to defend against direct oxidant injury to the lung and help to prevent inactivation of the elastase-inhibiting capacity of alpha-proteinase inhibitor (Galdston et al., 1984).

Other antioxidants have been found in alveolar fluid, including vitamin E (alpha-tocopherol), vitamin C (ascorbic acid), glutathione, and a catalase-like antioxidant. Studies by Pacht et al. (1986) evaluated the concentrations of vitamin E in the alveolar fluid of smokers and nonsmokers with BAL. The smokers' alveolar fluid had less vitamin E (3.1 $\pm$ 0.7 ng/ml vs. 20.7 $\pm$ 2.4 ng/ml). Vitamin E might be an important lower respiratory tract antioxidant, and its deficiency in smokers might predispose their lung parenchymal cells to oxidant injury.

There is also evidence that the extracellular fluid lining of the alveolus is enriched in endogenous ascorbic acid (Willis and Kratzing, 1976; Snyder et al., 1983). By direct measurement of total ascorbate in the lung lavage fluid of rats, Snyder et al. (1983) estimated the ascorbate concentration of alveolar lining fluid at 6.3 mmol/L, which is considerably higher than the 0.1 mmol/L in normal rat serum and high enough to allow the ascorbate to be an effective antioxidant. Studies at the Health Effects Research Laboratories of the Environmental Protection Agency are measuring the ascorbic acid content of nasal and pulmonary lavage fluids from control and ozone-exposed people (Hatch et al., 1987). Such studies should provide valuable insight into the role of extracellular antioxidants in protecting the lung from oxidant stress.

Reduced glutathione (GSH) is a sulfhydryl-containing tripeptide that is known to play an essential role in the defense of lung parenchymal cells and in the detoxification of various pollutants. Suther-

land et al. (1985) reported the presence of significant amounts of GSH in rat alveolar lining fluid, and Sun et al. (1985) reported GSH in BAL fluid from CD-1 mice equal to 50% of total GSH in lung tissue. Henderson et al. (1985a) reported increased GSH in BAL fluid from rats and mice exposed chronically to diesel exhaust. Studies by Cantin et al. (1985) have found that GSH is present in human alveolar lining fluid at concentrations far greater than those in plasma (approximately 40 μM vs. approximately 3 μM). Such concentrations of GSH are sufficient to protect lung parenchymal cells against H_2O_2 and suggest that GSH is an important extracellular antioxidant in the lower respiratory tract. Cantin and Crystal (1985) identified a catalase-like antioxidant in epithelial lining fluid that could play a role in defending parenchymal cells from oxidant injury. Taken together, those studies suggest that antioxidant components of alveolar lining fluid provide a first line of defense against oxidants in the lower respiratory tract.

Arachidonic Acid Metabolites

The synthesis and function of arachidonic acid metabolites remain the focus of extensive investigation (Gordon et al., 1988; Regal, 1988). The active products of membrane-derived arachidonic acid have remarkable biologic potency. However, the regulation of this metabolic system is still poorly understood.

Arachidonic acid is converted to pharmacologically active substances through two major synthetic pathways. The cyclo-oxygenase pathway generates unstable endoperoxides from which the primary prostaglandins—PGD_2, PGE_2 (dinoprostone), PGF_2, PGI_2 (prostacyclin)—and thromboxane A_2 are derived. The lipoxygenase pathway leads to the formation of hydroperoxy-eicosatetraenoic acid (HPETE), from which the leukotrienes—LTB_4, LTC_4, LTD_4, LTE_4—and the HETEs (5,12,15-lipoxygenases) are believed to be derived. In the lung, the major sources of cyclo-oxygenase products are AMs, fibroblasts, smooth muscle cells, and type II epithelial cells; endothelial cells are rich sources of prostacyclin, and platelets are active producers of thromboxane A_2. The most important cellular sites of lipoxygenase reactions are mast cells, basophils, and neutrophils (Said, 1982; Lewis and Austen, 1984; Lewis, 1985). Human airway epithelial cells can also selectively generate 15-lipoxygenase metabolites of arachidonic acid (Hunter et al., 1985).

Generally, prostaglandins of the D and F series are vasoconstrictors, whereas those of the E series and PGI_2 are vasodilators. PGE_2 acts as a vasodilator in the fetus and newborn and a weak vasoconstrictor in the adult. Prostaglandins, unlike leukotrienes, neither increase vascular permeability nor promote chemotaxis.

The principal products of the lipoxygenase pathway are leukotrienes. Three (LTC_4, LTD_4, and LTE_4) are sulfidopeptide leukotrienes and constitute the mediator originally described as slow-reacting substance of anaphylaxis (SRS-A). The compounds induce a sustained bronchospasm that is greater in peripheral than in central airways and can play a role in bronchial asthma (Weiss et al., 1982). It has been suggested that they can cause bronchial hyperresponsiveness that characterizes the asthmatic condition (Griffin et al., 1983). However, further research is needed to elucidate their exact role. In addition to their putative role in asthma, they have been shown to mediate increased vascular permeability in some tissues (Lewis and Austen, 1984), but their role in the lungs is uncertain. Studies with LTB_4 have shown it to be a potent chemotactic agent that produces endothelial cell adherence.

Arachidonic acid metabolites are not stored in tissues, but are synthesized de novo in response to stimuli. Their role as mediators of lung disease is only beginning to unfold. With the advent of sensitive radioimmunoassay and high-performance liquid chromatographic (HPLC) techniques, it is now possible to measure them in biologic fluids and study their production in isolated cultured cell systems. A number of recent studies have examined the presence of those compounds in BAL fluid.

Murray et al. (1986) performed BAL on five patients with chronic stable asthma before and after local challenge with *Dermatophagoides pteronyssimus*. The BAL fluid was analyzed for arachidonic acid metabolites. In the five patients, PGD_2 concentrations increased by an average of a factor of 150 after local instillation of antigen. The results constitute evidence that the release of PGD_2 into the airways is an early event after the instillation of *D. pteronyssimus* in patients who are sensitive to this antigen.

The oxidant air pollutant ozone can produce airway inflammation and hyperresponsiveness in exposed people. Seltzer et al. (1986) exposed 10 healthy human subjects to air or O_3 (0.4 or 0.6 ppm). Airway responsiveness to inhaled methacholine was measured before and after each exposure, and BAL was performed 3 hours after the exposure. An increase in the number of neutrophils was found in BAL fluid from O_3-exposed subjects, especially those in whom O_3 exposure produced an increase in airway responsiveness. They also found significant increases in PGE_2, PGF_2, and thromboxane B_2 in BAL fluid from O_3-exposed subjects. Hence, O_3-induced hyperresponsiveness appears to be associated with both neutrophil influx and changes in the concentrations of some cyclo-oxygenase metabolites.

Other studies (Laviolette et al., 1981) have examined the production of arachidonic acid metabolites by human AMs recovered from smokers and nonsmokers. PGE_2 and thromboxane B_2 synthesis was significantly lower in AMs from smokers than in those from nonsmokers. A cigarette-smoke-induced lesion in phospholipid hydrolysis is most consistent with the findings. Inasmuch as arachidonic acid metabolites are involved in the regulation of immune and inflammatory responses and bronchiolar and vascular smooth muscle reactivities in the lung, it was concluded that the defect observed in smokers' AM can play a role in the pathogenesis of cigarette-smoke-induced diseases.

Animal studies have also attempted to correlate exposure to airborne substances with increases in lavage-fluid arachidonic acid metabolites. Mundie et al. (1985) exposed New Zealand white rabbits to aerosolized cotton dust extract and performed lavage at various times after exposure. PGF_2, PGE_2, and thromboxane B_2 were maximally increased in the lavage fluid 4 hours after exposure. Results of that study, the first to demonstrate in vivo release of arachidonic acid metabolites in the lung in response to inhalation of cotton dust extract, strongly suggest that the metabolites are responsible for the bronchoconstriction seen in the acute byssinotic reaction in humans. Further studies that establish the presence of those metabolites at sites of lung injury are needed, for their role in the pathogenesis of lung injury to be understood.

Complement

The complement system, which is composed of more than 20 plasma proteins, is an important mediator of various inflammatory responses, such as increase in vascular permeability and chemotaxis of PMNs (Colten et al., 1981). The system comprises two major pathways—classical and alternative (Perez, 1984). The classical pathway involves the union of antigen and antibody, which then binds to and activates the C1 complex. The alternative pathway involves direct contact of the subject with antigens, such as bacteria, fungi, endotoxins, immune complexes, and particles (Wilson et al., 1977; Warheit et al., 1985, 1986).

Active products of the complement system, such as C3b, have been shown to promote phagocytosis. C3a and C5a, however, can cause mast cells to degranulate and release histamine, and they can increase vascular permeability. C5a is also chemotactic for PMNs and promotes the release of their lysosomal enzymes (Perez, 1984).

BAL fluid from normal persons contains components of both the classical and alternative pathways of complement, i.e., C3, C4, C5, C6, and factor B (Reynolds and Newball, 1974; Robertson et al., 1976; Henson et al., 1979). Although some of those components can be derived via transudation from serum, local production by lung fibroblasts and AMs is also possible (Colten and Einstein, 1976; Reid and Solomon, 1977; Cole et al., 1983). During

inflammation, when vascular and epithelial membrane permeability is increased, additional complement components of plasma can enter alveolar spaces and airways.

If it is feasible to use enumeration and analysis of inhaled particles in situ as a marker of exposure, then it should be feasible to use cellular responses to such exposure as further markers of exposure, as well as predictors of injury at the alveolar level. At least two biologic responses take place very rapidly after exposure to a variety of inorganic particles. The first is activation of the fifth component of complement, C5 (Warheit et al., 1985), and the second is macrophage accumulation, which is a consequence of the activation (Warheit et al., 1984-1986).

Several complement components normally are found in the complex alveolar lining layer. The role of complement on alveolar surfaces is not entirely clear, but one important function appears to be the clearance of inhaled microbes (Larsen et al., 1982). It has been established that C5 can be activated through the alternative pathway to produce C5a, a potent chemoattractant of neutrophils and macrophages (Snyderman, 1981). It was recently shown that the C5 on alveolar surfaces was activated by a variety of inhaled particles (Warheit et al., 1988) chrysotile asbestos being a noteworthy example. During a 3-hour exposure to asbestos, all detectable C5 on alveolar surfaces was converted to C5a (Warheit et al., 1986). C5a, a chemotactic factor, remained active in the alveoli for about a week, and concentrations of C5 returned to normal 1-2 weeks after the 3-hour exposure (Warheit et al., 1986).

Could that biologic response be used as a marker of exposure to inorganic particles, as described in Chapter 2? Alveolar fluids collected with BAL have been separated by appropriate biochemical techniques. If the normal extent of complement-dependent chemotactic activity were established, activation of C5 might well serve as a marker of exposure. The cellular response to activation should also be predictable and could serve as an additional, correlated marker.

Activated C5, whether from serum complement or from alveolar complement, attracts neutrophils and macrophages (Warheit et al., 1986). Several studies have shown that it is possible to predict whether macrophages will be attracted to specific bacteria or inorganic particles on the basis of their capacity to activate C5 in vitro (Warheit et al., 1988). For example, chrysotile asbestos and crocidolite asbestos are good activators of C5; after inhalation, they attract macrophages to alveolar duct bifurcations. However, ash from Mount Saint Helens induces no detectable C5a production and attracts few macrophages that have been stimulated to migrate by C5a. It is conceivable that such cells, easily recovered from the lung, could serve as markers of exposure if their biology were better understood.

There is a vast literature on macrophage physiology and function aimed at developing a better understanding of macrophages (Fels and Cohn, 1986). Those cells are avid phagocytes, so determination of their particle burden has proved to be an extremely useful marker of exposure (Brody, 1984). However, in humans, recovery of macrophages from the lung is usually too late to yield an early marker of exposure. It might therefore be important to consider initial complement activation and the later macrophage response in estimating exposure in humans.

Once the macrophages have responded to inhaled agents, it is reasonable to conclude that the cells will release a variety of products (Fels and Cohn, 1986), many of which are known to have profound effects on pulmonary cells and tissues. Three well-known examples of such products are oxygen radicals, arachidonic acid metabolites, and growth factors for fibroblasts. Whether any or all of them could serve as useful markers of exposure and injury is yet to be determined. But it is reasonable to suggest that they all could become markers of pulmonary insult.

Recently, investigators have begun to look at complement activity in BAL fluid from patients with pulmonary disease. For example, Lambré et al. (1986) demonstrated the presence of C3b and Bb (the activated forms of the proteins C3 and B) in BAL fluid from patients with pulmonary

sarcoidosis. Complement activity in lavage fluid decreased in patients receiving corticosteroid therapy. That suggests that complement activity in the alveolar spaces might be a good marker of the activity of the disease in lasting sarcoidosis. There have also been reports of the presence of C3b in lavage fluid from patients with idiopathic pulmonary fibrosis (Robbins et al., 1981). Results of such studies suggest that the complement system can play a role in the pathogenesis of those diseases. Further studies are needed, however, to validate the use of complement activity as a marker of lung injury and disease.

Growth Factors and Monokines

In addition to their role as the primary phagocytes in the lung, AMs synthesize diverse substances that exhibit a broad range of biologic activities, including mediators (monokines) that regulate the growth or activation of other cells. Pulmonary AMs, on activation, release two primary growth factors for lung fibroblasts: fibronectin and AM-derived growth factor, or AMDGF (Bitterman et al., 1986). Fibronectin, a 440,000-dalton glycoprotein, has been shown to act as a "competence factor"; it delivers a growth-promoting signal to nonreplicating lung fibroblasts early in the G1 phase of the cell cycle. AMDGF, an 18,000-dalton peptide, provides the second (progression signal) of two required signals in G1 to induce fibroblasts to divide. Although fibronectin is a normal constituent of the alveolar epithelial lining fluid, Rennard and Crystal (1982) have shown that its concentration is 2-5 times higher in patients with fibrotic lung disorders. In fact, pulmonary AMs from most patients with interstitial fibrosis have been shown to release both fibronectin and AMDGF (Bitterman et al., 1986). However, the exact role of these growth-modulating signals in the pathogenesis of chronic interstitial disorders remains to be elucidated.

AMs release interleukin 1 (Il-1), a monokine that is a lymphocyte-activating factor, in response to various immune or inflammatory stimuli. Il-1, a protein of

12,000-18,000 daltons, is thought to be important in modulating T- and B-cell activation and in other inflammatory processes (Wewers et al., 1984).

Many of the substances that induce the secretion of Il-1 can also stimulate AMs to secrete interferon-gamma. Interferon-gamma augments T-cell replication, apparently by inducing Il-2 receptor expression (Johnson and Farrar, 1983).

Clearly, stimulated AMs can release a wide array of potent biologic mediators. However, factors governing the selective release of those mediators into the lower respiratory tract are still poorly understood.

Oxygen Radicals

Considerable evidence accumulated in recent years suggests that oxygen-derived free radicals are an important cause of tissue injury in many disease processes (Freeman and Crapo, 1982). The lung is prone to oxidant stress with a variety of sources, and it has been suggested that a reactive oxygen species plays a role in the development of acute lung injury and the etiology of chronic lung disease (Johnson et al., 1981).

As described previously, oxidants can be generated in the lower respiratory tract via the action of PMNs and AMs involved in local inflammatory reactions. On recognition of a phagocytic or soluble stimulus, both neutrophils and macrophages experience a "respiratory burst" that is characterized by an increase in oxygen consumption, activation of the hexose monophosphate shunt, and the generation of reactive oxygen species, including O_2^-, H_2O_2, and $OH^\bullet$. That burst of activity is related to the stimulation of membrane-bound reduced nicotinamide adenine dinucleotide phosphate (NADPH) oxidase (Babior, 1978). Those oxygen-derived products normally play a major role in phagocyte-mediated bactericidal activity, but it is conceivable that they contribute to host tissue injury when their production is stimulated inappropriately. A number of models that attempt to assess the effects of oxygen radicals on the lung have been developed.

Johnson et al. (1981) demonstrated that acute lung injury occurs if specific enzyme-substrate systems known to generate oxygen metabolites are intratracheally instilled into rat lungs. For example, if xanthine and xanthine oxidase (which generate O_2^-) was administered intratracheally, there was increased vascular permeability with minor edema formation and focal hemorrhage after 4 hours. Those pathologic changes could be inhibited by simultaneous instillation of superoxide dismutase (SOD). However, when glucose and glucose oxidase (which generate H_2O_2) were instilled into the airways, there was a marked increase in vascular permeability, edema, hyaline membrane formation, hemorrhage, and neutrophil influx. Those changes are consistent with the human pathologic changes referred to as diffuse alveolar damage and associated with adult respiratory distress syndrome (ARDS). The changes could be inhibited with catalase, but not SOD. Furthermore, if either lactoperoxidase (LPO) or myeloperoxidase (MPO) was instilled with the glucose-glucose oxidase system, severe lung injury occurred and frequently progressed to diffuse pulmonary fibrosis by 4 days. The data suggest that a product of MPO (or LPO), H_2O_2, and halide (perhaps HOCl) plays an important role in the development of pulmonary fibrosis. Oxygen-derived free radicals and their metabolites can cause acute lung injury and progressive lung injury with pulmonary fibrosis.

In another series of experiments, Johnson and Ward (1982) instilled phorbol myristate acetate (PMA), a potent initiator of the respiratory burst, intratracheally into neutrophil-depleted rats. They found that the instillation caused acute lung injury that was inhibited by catalase, but not by SOD; again, H_2O_2 was implicated as the cause of the damage. The source of the toxic H_2O_2 in the model appears to be PMA-stimulated AMs. Some investigators have recently found that AMs retrieved from some cigarette-smokers spontaneously release H_2O_2 (Greening and Lowrie, 1983; Baughman et al., 1986b). That suggests that cigarette smoke can activate these cells (Hoidal and Niewoehner, 1982). Although activated phagocytes can release substantial amounts of potent oxidants to surrounding tissues, they are by no means the only source of reactive oxygen species in the lung. Reduction of O_2 to active O_2 metabolites occurs as a byproduct of cellular metabolism during microsomal and mitochondrial electron transfer reactions (Cohen and Cederbaum, 1979); considerable amounts of O_2^- are generated by NADPH-cytochrome P-450 reductase reactions (Kameda et al., 1979). Because those metabolites are potentially cytotoxic, they might mediate or promote actions of various pneumotoxins.

Such mechanisms have been proposed for paraquat- and nitrofurantoin-induced lung injury (Sasame and Boyd, 1979; Shu et al., 1979). Similarly, Freeman and Crapo (1981) demonstrated that hyperoxia increases the steady-state concentrations of O_2^- and H_2O_2 in lung tissue and that mitrochodria contribute importantly to this phenomenon. Those observations lend support to the hypothesis that lung damage during hyperoxia is mediated by increased production of oxygen radicals.

In conclusion, reactive oxygen species in the lung can be generated by multiple and diverse processes and appear to play a role in the onset of acute lung injury and possibly in the development of chronic lung disease. Additional studies are necessary, to define the precise targets of oxygen metabolites in the lung and the specific biochemical mechanisms by which the oxidants damage lung cells.

Enzymes

Increases in enzymatic activities in BAL fluid have been used as markers of pulmonary responses to inhaled toxicants, particularly in animals (Beck et al., 1982; Henderson, 1988a,b). Extracellular lactate dehydrogenase (LDH), a cytoplasmic enzyme, is used as a marker of cytotoxicity, because LDH is not found extracellularly except in the presence of damaged or lysed cells. This marker has been used in numerous studies, for example, in hamsters exposed to mineral dusts (Beck et al., 1982), in rats and mice exposed to diesel exhaust (Henderson et al., 1988), and in sheep exposed to asbestos (Begin et al., 1983). Another cytoplasmic enzyme that

has been assayed in BAL fluid is glutathione reductase (Henderson et al., 1988).

Lysosomal enzymes, such as N-acetylglucosaminidase (Beck et al., 1982) and beta-glucuronidase (Henderson, 1988b), appear to be good indicators of increased phagocytic activity in response to inhaled particles. The extent of the increases in BAL-fluid lysosomal enzyme activities appears to correspond to the toxicity of the inhaled particles (Beck et al., 1982; Henderson et al., 1985a) and exceeds the degree of increase in LDH by several fold (Henderson et al., 1985b). The increase in lysosomal enzymes relative to LDH can be used to estimate how much of the increase in lysosomal enzymes is due to lysed cells (which would cause concomitant release of lysosomal enzymes and LDH) and how much is due to stimulated phagocytic cells. Acid phosphatase, also a lysosomal enzyme, does not increase in BAL fluid in response to inhaled particles (Henderson et al., 1985b). Either that enzyme is not in the same lysosomal storage site as beta-glucuronidase and similar hydrolytic enzymes or it is rapidly broken down in the epithelial lining fluid, once released.

Increases in alkaline phosphatase activity have been detected in BAL fluid from NO_2-exposed hamsters (DeNicola et al., 1981). A lung-specific form of this enzyme has been reported to be released from Type II pneumocytes (Reasor et al., 1978; Miller et al., 1986). A histochemical stain specific for the enzyme has been used as a marker of Type II cell proliferation (B. E. Miller et al., 1987).

Proteolytic activity and antiproteolytic activity in BAL fluid are of interest, because an imbalance between the two could lead to breakdown of lung tissue, such as that seen in emphysema (Janoff, 1972; Starkey and Barrett, 1977). Proteolytic enzymes detected in BAL fluid include collagenase, PMN elastase, metalloproteinase, plasminogen activator, and acid proteinases (Barrett, 1977a,b; Harper, 1980; Gadek et al., 1980; Pickrell, 1981). Antiproteinases in BAL fluid are alpha$_1$-antiproteinase, alpha$_2$-macroglobulins, and bronchial antiproteinase. Acid proteinase activity is associated with lysosomes and is released with other lysosomal enzymes in response to inhaled toxic particles (Wolff et al., 1988).

Protein and Protein Products

Protein in BAL fluid is measured as a marker of increased permeability of the alveolar-capillary barrier and is a common component of the inflammatory response. Bell and Hook (1979) reported that 80% of the soluble protein in human BAL fluid could be accounted for by 19 plasma proteins. The protein content indicated a preferential transfer of smaller proteins across the alveolar-capillary barrier. IgG and IgA constituted a higher fraction of total protein in BAL fluid from smokers than in serum (Bell et al., 1981). Transferrin was the only nonimmunoglobulin protein with a higher concentration in lavage fluid than in serum. Serum proteins in BAL fluid from animal studies have proved to be sensitive markers of the inflammatory response (Alpert et al., 1971; Bignon et al., 1975; DeNicola et al., 1981; Beck et al., 1982; Lehnert et al., 1986).

The amino acid hydroxyproline is a marker of collagen and has been interpreted as a marker of collagen breakdown. Hydroxyproline content of BAL fluid has been measured as a marker of breakdown or remodeling of pulmonary collagen in ozone-exposed rats (Pickrell et al., 1987). The increase in hydroxyproline in BAL fluid appeared to parallel developing pulmonary fibrosis in hamsters and rats exposed to diesel exhaust (Heinrich et al., 1986; Henderson et al., 1988).

MOLECULAR MARKERS

Exposure to environmental toxicants can cause damage in single cells at the level of DNA, and that damage can lead to the development of many diseases, including cancer. Toxicant-induced changes in specific (although often unidentified) genes are thought to be the initial events in the development of disease. Identification of genes involved in the development of specific diseases can lead to improved diagnosis, understanding, and treatment, but is not essential. In lieu of disease-specific molecular markers that could be used to study the relationship between toxicant exposure and the development of disease, the general interaction between toxicants and DNA can serve as a source of molecular markers of exposure, effect, and susceptibility. The use of molecular markers, defined here as alterations in DNA or RNA, to identify cellular responses or responsiveness to environmental toxicants theoretically can provide information useful in determining the magnitude of exposure, the effects of exposure on human health, and the mechanisms of response. This section discusses some general considerations in the use of molecular markers, defines some general types of molecular markers, identifies specific markers for potential use in pulmonary toxicology or the study of carcinogenesis, and identifies subjects for research that could lead to the identification of new molecular markers.

Molecular markers can be highly sensitive and specific indicators of cell damage or change. Detection of toxicant-induced alterations and use of them as indicators of toxicant exposure, effect, or susceptibility depend on several factors, including the frequency of the alteration, which in turn can affect the sample size required for its detection; the availability of sufficient material (DNA, RNA, or cells) for analysis; and the accessibility of the cells at risk (can they be obtained noninvasively, or are invasive procedures required?).

The sample size required for detection of toxicant-induced alterations is a major consideration in the choice or use of a marker. The minimal sample size required for a given assay depends on the sensitivity of the assay and on the fraction of cells in a sample that contain the specific change of interest. Changes found in a large fraction of cells in a sample will be detectable with a much smaller sample than changes found in only few cells in a sample. For example, many assays that involve the analysis of a DNA change require about 5-10 μg of DNA from cells containing the change of interest. That amount of DNA can be obtained from 10^6 altered cells. Obtaining 10^6 altered cells might require a sample of as few as 10^6

cells, if the change occurred in all cells after exposure or if the cells being used all came from a specific exposure-induced lesion. But a sample of 10^{12} cells could be required, if the change occurred with a frequency of 10^6, which is the observed frequency of induction of some single-gene mutations (Baker et al., 1974). A sample of 10^6 cells is readily obtainable with BAL or even with a small tissue biopsy, but a sample of 10^{12} cells is more difficult to obtain.

In addition to the frequency of the change under investigation, accessibility and availability of sample material affect the choice and use of an assay. Common changes found in cells in BAL fluid, because of their greater accessibility and availability, are much easier to detect than changes (even common ones) that occur only in cells of the deep lung. Assays that depend on invasive sampling procedures can be useful if discrete lesions are being biopsied. However, the routine use of invasive sampling procedures before a lesion is identified usually cannot be justified.

The use of molecular markers as indicators of exposure might therefore be limited to cases in which changes are of a general nature, in which changes occur in readily accessible cells, or in which discrete exposure-induced lesions are being biopsied. Changes that are rare or cell-specific can be difficult or impractical to detect if large tissue samples or invasive sampling procedures are required. However, molecular markers potentially can play an important role in mechanistic studies of disease.

Potential molecular markers can be divided into several categories, including those based on genetics (modifications of DNA bases, changes in DNA sequence or structure, and changes in extent or pattern of gene expression) and those based on their ability to detect toxicant exposure, effect, or susceptibility or their ability to identify the toxicant involved.

Markers of toxicant exposure could be used as screens for exposure to a given toxicant. Depending on the assay and the markers involved, markers theoretically could be used to indicate simply that ex-

posure to a toxicant occurred, to estimate the extent of exposure, or to identify the toxicant. Markers of exposure should be readily detectable and measurable in an accessible population of cells. In addition, toxicant-induced changes should be detectable soon after exposure, and the persistence of a given marker should be determined. Finally, for a marker to be useful as an indicator of exposure to a specific toxicant, it should be characterized sufficiently for its presence to be attributed to a given toxicant with reasonable certainty. Molecular markers could also be used to study the biologic effects of exposure to specific toxicants. They could be used to monitor or characterize the development of toxicant-specific responses, such as alterations in gene expression. Such analyses would permit studies in the early stages of response, before the development of toxicant-induced lesions or disease. Molecular markers theoretically could be used to identify people with an increased risk of the effects of particular toxicants; that would make it possible to minimize their exposures.

The formation of DNA adducts after exposure to chemicals is an example of an exposure-related modification of DNA (Poirier and Beland, 1986a; NRC, 1989). DNA adducts form when chemicals or metabolically activated derivatives of them bind covalently to DNA. The presence of adducts can be detected chemically (Belinsky and Anderson, 1987; Gupta, 1987) or immunologically (Santella et al., 1987). The most sensitive assay for the detection of DNA adducts is the ^{32}P postlabeling assay (Gupta, 1987). For maximal sensitivity, the assay requires 5-10 μg of DNA (Gupta, 1987), which can be obtained from a sample of 10^6 cells. Sufficient cells are generally available for this assay, because DNA adducts can be found in cells in a variety of readily available biologic samples, such as blood (e.g., adducts of lymphocyte DNA) and BAL fluid (e.g., adducts of macrophage DNA), depending on the type of exposure.

DNA adducts are more useful as indicators of exposure or dose than as markers of effect; sequence-specific or gene-specific adduct formation has yet to be demonstrated. However, care must be taken even in the use of adduct concentrations as indicators of total exposure, in that the concentrations in tissues can be affected by a variety of biologic responses, including adduct repair and cell turnover, which vary from one tissue to another (Belinsky and Anderson, 1987).

A correlation between the concentration of DNA adducts or the presence of specific adducts and the development of disease remains to be proved. The presence of a common adduct induced by a particular treatment might have little biologic consequence, whereas the presence of a rare adduct induced by the same treatment could be highly significant (Poirier and Beland, 1986b). Identification of exposure-specific adducts and demonstration of an association between specific adducts and toxicant-induced disease will expand the use of adducts from indicators of exposure and estimators of total dose to specific tools for identifying toxicants and estimating risk of disease.

Another type of DNA modification that could be affected by cell responses to various exposures is DNA methylation. Changes in patterns of DNA methylation potentially could be used as indicators of cellular response or of cellular responsiveness to particular toxicants. Site-specific changes in the extent of DNA methylation have been shown to regulate gene expression (Razin and Riggs, 1980; Feinberg and Vogelstein, 1983). Some toxicants could result in changes in DNA methylation patterns and cause exposure-related alterations in gene expression. Changes in DNA methylation can be detected either as changes in gene expression or as changes in the restriction-enzyme sensitivity of the genes involved, because of altered methylation of restriction-enzyme recognition sequences (Razin and Riggs, 1980). Identification of changes in DNA methylation requires identification of the affected gene(s) and analysis of the methylation changes in the gene(s), in that alterations in the extent of DNA methylation at the whole-cell level are difficult if not impossible to detect. Furthermore, large numbers of cells (more

than 10^6) containing the same changes in methylation would be needed, so biopsies of developing lesions would usually be required. That approach is more likely to be useful in retrospective studies of mechanisms of cellular response than as a source of markers of response.

Changes in DNA sequence or structure could be a source of exposure-related molecular markers. Structural damage to DNA, such as double-strand or single-strand breaks, can be detected with filter elution assays and alkaline or neutral elution (Bradley et al., 1982). Because those assays, like the assay for DNA adducts, detect general cell damage, a random sample of 10^6 cells would provide enough DNA for analysis. Structural DNA damage can also be detected by examining exposed cells for chromosomal aberrations. After exposure to some toxicants, chromosomal aberrations have been detected in macrophages isolated by BAL (Au et al., 1988). The lack of gene or sequence specificity of the assays makes them most useful as indicators of exposure to particular toxicants or as estimators of total dose.

Changes in DNA sequence resulting from point mutations or deletions are likely to be the initiating events of some exposure-related cellular responses, such as tumor development. One indicator of those changes at the cellular level is the production of mutations after exposure. Many toxicants have been shown to be mutagens in assays that use mammalian cells or bacteria in vitro (Ashby, 1982). One means of measuring in vivo mutations in man uses lymphocytes and changes in the hypoxanthine-guanine phosphoribosyl transferase (HPRT) gene (Albertini, 1980). Cells that are deficient in HPRT can proliferate in the presence of the toxic purine analogue 6-thioguanine, whereas HPRT-normal cells cannot. HPRT mutants have been detected (with autoradiography) by their ability to form colonies in a 6-thioguanine medium (Morley et al., 1983) or by their incorporation of [^{3}H]thymidine in the presence of 6-thioguanine (Albertini, 1980). Lymphocytes from persons exposed to toxicants could be examined for 6-thioguanine resistance, although measurable increases in the frequency of mutants would be expected only in cases that resulted in systemic exposure to toxicants or their metabolites. Alternatively, the assay could be adapted for use with macrophages isolated by BAL. The use of toxicant-induced mutations as markers of DNA damage would provide information on exposure and total dose.

The identification of toxicant-induced changes in DNA sequence at the molecular (as opposed to cellular) level is important in understanding the etiology of some toxicant-induced diseases, but the changes are not likely to be a useful source of markers of toxicant exposure or of early stages of disease. Detection of changes in the base sequence of specific genes requires that the altered DNA be isolated and examined, with radiolabeled molecular probes, for specific changes in DNA sequence (Reddy et al., 1982). The sensitivity limits of that type of assay require the presence of a minimum of 1 picogram (pg) of DNA with the sequence of interest (Thomas, 1983). For example, if the gene of interest were a single-copy gene encoded by 5,000 base pairs of DNA, 1 pg of DNA from the altered gene could be obtained from a minimum of 2×10^5 altered cells. Detection of point mutations within the sequence of a particular gene generally requires that the gene be cut into multiple fragments with restriction enzymes for analysis by gel electrophoresis, so up to 10^6 cells might be required to yield 1 pg of DNA with the sequence of interest. As noted above, isolation of so many cells from an exposure-induced lesion by noninvasive methods is not likely. Sampling of cells with lavage will yield more than 10^6 total cells, but most of the cells will not contain the change of interest. That approach is most likely to be useful in retrospective analyses of mechanism, not in surveys of exposure effect.

Changes in the amount or pattern of gene expression that result from exposure to some toxicants might be most amenable to the use of molecular analysis as a measure of effect. If the expression of specific (identified) genes is induced, amounts of mRNA in the target cells might be greatly increased (or reduced). The detection and measurement of mRNA by Northern or dot blot analysis requires the presence of 1 pg of the sequence of interest (Thomas, 1983). However, the expression of a specific gene can result in the production of large amounts of mRNA for the gene of interest; that decreases the number of cells required for detection. For example, induction of the ovalbumin gene in the oviduct gland cell of chickens results in the production of more than 3,000 ovalbumin mRNA fragments per cell, compared with the noninduced number of 2 copies per cell (Roop et al., 1978). That amplification (by a factor of 1,500) reduces the number of cells required to obtain 1 pg of ovalbumin mRNA from about 5×10^5 to about 300. A gene- and exposure-specific response could therefore be followed with molecular markers if specific (identified) genes were overexpressed after exposure to particular toxicants, if molecular probes for the genes were available (i.e., if the genes had been isolated and molecularly cloned), and if the cells containing the overexpressed genes were readily available in sufficient numbers (e.g., in lavage fluid) from exposed persons.

If molecular probes for specific genes of interest were not available for use in the assay described above, exposure- or disease-specific changes in gene expression could be examined, provided that assays for the gene product(s) were available. Poly(A)-containing RNAs isolated from affected tissues could be translated in vitro with a reticulocyte lysate system (El-Dorry et al., 1982), and the amount and nature of protein product could be analyzed to detect exposure- or disease-related changes. The assays for alterations in gene expression, although potentially useful in understanding the mechanism of response to a toxicant once a response has occurred, are not likely to be useful for surveys of exposure or effect.

Gene-specific changes, such as specific sequence changes or modifications of DNA, generally occur too infrequently to be useful as markers of toxicant exposure and often occur in cells accessible only with invasive sampling procedures once lesions have been identified. However, some gene-specific changes could be useful as general markers of toxicant exposure.

For example, exposures that result in inflammation involve recruitment and activation of macrophages that express genes for interleukin-1 and c-sis (Wewers et al., 1984; Mornex et al., 1986). Molecular probes are available for those genes, so their expression or changes in their expression could be detected with mRNA isolated from macrophages in lavage fluid. Changes in the expression of the genes could serve as general indicators of exposure. The exposure-related changes in gene expression would also provide information useful in understanding the mechanism of toxicant effects.

Available markers potentially could be used to detect specific exposure-induced effects. The ability to detect changes at different stages of disease will depend on the availability of sufficient cells for analysis. Specific exposure-induced effects might be detectable at the molecular level only at more advanced stages of disease and are likely to be more useful in the characterization of a disease than in its diagnosis. Genes for pulmonary surfactant apoprotein (White et al., 1985), for collagen (Miskulin et al., 1986), and for cytochrome P-450 enzymes involved in oxidative metabolism (Nebert and Gonzalez, 1987) have been cloned. Changes in the expression or molecular structure of those genes after toxicant exposure could be identified. Exposures—such as chronic exposure to cigarette smoke—that affect Type II cells result in alterations in surfactant production (Le Mesurier et al., 1981). Those changes could be characterized at the molecular level with available probes. Similarly, the induction and expression or overexpression of genes for collagen could be examined after exposures that result in excess collagen deposition. Finally, the toxicant-specific induction and expression of genes for cytochromes P-450 could be monitored with cloned probes. The success or feasibility of each of those analyses is subject to the same restriction of cell availability as described above.

The analysis of cancer development after toxicant exposure is another endeavor in which molecular probes could be useful for understanding the mechanism of response and possibly as a diagnostic tool. Alterations in the number of copies or expression of cellular oncogenes have been identified in several pulmonary cancers. For example, amplification of *c-myc* has been found at a late stage in the development of some small-cell lung carcinomas (Little et al., 1983; Saksella et al., 1985), mutationally activated *K-ras* has been found in some lung carcinomas (Santos et al., 1984; Stowers et al., 1987), and overexpression of *erb-B* has been described in some non-small-cell lung carcinomas (Cerny et al., 1986; Gamou et al., 1987). Theoretically, exfoliated tumor cells could be identified and characterized from lavage fluid with in situ hybridization, if tumor-specific oncogene changes were established. In addition, toxicant-specific oncogene activation could be characterized in developing lung tumors. Some examples of carcinogen-specific oncogene activation have been described. Induction of mammary tumors in rats with nitrosomethylurea resulted in *H-ras* activation in 86% of developing tumors (Zarbl et al., 1985). Similarly, 74% of lung tumors in rats exposed to tetranitromethane had an activated *K-ras* (Stowers et al., 1987). Those and other examples of carcinogen-specific oncogene activation suggest that analyses of oncogene activation in tumors after environmental exposures could play a role in increasing understanding of the etiology of exposure-related tumor development.

In conclusion, there is clearly a need for more markers that can be used to detect and characterize at the molecular level both general and specific cell responses to exposure. Studies at the molecular level will continue to be most useful in understanding mechanisms of cellular response to toxicants. However, it might be possible to develop specific molecular probes that could be used to diagnose or characterize specific diseases or other responses to exposure. Molecular probes have proved useful in the diagnosis and characterization of some infectious diseases and in sickle-cell anemia and alpha- and beta-thalassemia.

Molecular markers that could identify individual susceptibility to disease or toxicant sensitivity are also needed.

Molecular probes have been or are being developed for several diseases, including sickle-cell anemia and thalassemias (Dozy et al., 1979; Wilson et al., 1982; Pirastu et al., 1983), retinoblastoma (Friend et al., 1986), Huntington disease (Carlock et al., 1987), Duchenne muscular dystrophy (Monaco et al., 1986, 1987), cystic fibrosis (Dorin et al., 1987), Lesch-Nyhan syndrome (Brennand et al., 1982), phenylketonuria (Woo et al., 1983), antithrombin III deficiency (Prochownik et al., 1983), and alpha$_1$-antitrypsin deficiency (Kidd et al., 1983). Results of studies of chronic obstructive pulmonary diseases suggest a genetic basis (Kauffmann, 1984) and might therefore lead to the discovery of molecular markers of susceptibility. An increased risk of cigarette-smoking-induced bronchiogenic carcinoma appears to be associated with a highly inducible cytochrome P-450 phenotype (Jaiswal et al., 1985; Gonzalez et al., 1986). Further correlations between that phenotype and development of other pulmonary diseases are needed. Molecular analyses of other pulmonary diseases or individual responsiveness to toxicants might enable identification of persons at greater risk of developing toxicant-specific diseases and lead to the development of markers of susceptibility.

The development and use of molecular markers to identify cellular responses or responsiveness to environmental toxicants and to characterize pulmonary disease will be important in increasing understanding of the mechanisms involved in the development of pulmonary disease and in its prevention and treatment.

7

Conclusions and Recommendations

The potential health effects of exposure to environmental pollutants constitute a problem of great concern, because of the quantities of pollutants in question and the large numbers of people involved. But the problem is difficult to assess, because exposures to the pollutants and their mixtures are generally small. The use of biologic markers as objective measures of exposure or response to environmental pollutants offers a promising new approach to this problem. Biologic markers potentially can be used to obtain evidence of exposure to specific chemicals and of responses to exposure.

The study and use of biologic markers are growing rapidly. The list of macromolecular adducts formed because of exposures to chemicals is expanding daily. Since the drafting of this report itself, research on the functions of adhesive proteins, such as fibronectin and laminin, and of cellular integrins, which bind to receptor sites on these proteins, has exploded, providing potential new markers for following the progress of normal and abnormal processes of repair of pollutant-caused damage.

Listed below are recommendations for research that offers the best opportunity to enhance the use of biologic markers in the study of environmental health effects

on the respiratory tract. Recommendations aimed at increasing the use of biologic markers in the dosimetry of inhaled materials are listed first and are followed by recommendations regarding the use of such markers in detecting structural, physiologic, and biochemical responses to inhaled pollutants.

DOSIMETRY

1. More information is needed on factors that affect the dosimetry of inhaled toxins at specific sites along the respiratory tract. Specifically needed is information on:

• Regional deposition of inhaled pollutants at various sites along the respiratory tract. Considerable information is available on regional deposition of inhaled particles larger than .01 μm in aerodynamic diameter, but relatively little is known about factors that govern the deposition of inhaled gases, vapors and ultrafine particles (smaller than .01 μm in aerodynamic diameter). Specific factors that affect deposition, particularly airway structure, need to be assessed in both laboratory animals and humans.
• Pollutant effects on clearance of deposited material. Dosimetry at specific

sites in the respiratory tract depends both on how much is deposited at the site and on how quickly it is removed. Interspecies studies of regional clearance are required.

• The capacity of tissues at the site of deposition to metabolize a pollutant to a more or a less toxic form. The toxicity of an inhaled organic compound might be increased by metabolic activity at some sites of deposition.

• Effects of chemical and physical characteristics of pollutants on the site of sequestration and on the induction of injury in the respiratory tract.

2. Physiologic modeling of the pharmacokinetics of inhaled materials in animals and humans shows promise for allowing extrapolation of dosimetry data between species, sexes, and regimens. Extension of that approach to the active metabolites of inhaled compounds would greatly increase our understanding of the toxicity of inhaled materials. Physiologic modeling also requires information on deposition, clearance, and metabolism described in Recommendation 1 above.

3. One region of the respiratory tract that has received little attention but is readily accessible for sampling is the nose. The analysis of nasal rinses or sputum to detect exposures to specific pollutants could be useful. It must be applied with appropriate knowledge of dosimetry differences between the nose and the rest of the respiratory tract when different toxicants are inhaled.

4. Macromolecular adduct formation offers a promising new method of measuring exposure to organic compounds that are or can be metabolized to reactive forms. Further research on the kinetics of adduct formation and clearance is required to determine the relationship between exposure history and the concentration of adducts in blood or tissue samples.

5. Most research has been on formation of adducts with DNA and hemoglobin. Adducts with other macromolecules, particularly those with site specificity, should be explored as markers of dose.

PHYSIOLOGIC MEASUREMENTS

1. Existing physiologic tools need to be extended and new tests need to be developed and evaluated to focus on specific sites of action of environmental pollutants and specific effects of given pollutants. That requires evaluation of pathophysiologic correlates assessed initially in animals and later in humans, both in controlled exposure settings and in population-based samples.

2. Further research is required on the role of nonspecific airway reactivity in identifying persons susceptible to environmental agents and on the role of airway reactivity in the natural history of chronic obstructive pulmonary disease (COPD). The role of transient changes in airway reactivity in response to specific environmental agents needs to be assessed in regard to increased risk of development of COPD.

3. Linkages among altered particle clearance, environmental exposures and development of lung disease need to be studied further to determine the usefulness of clearance as a marker of susceptibility and response.

4. Markers of early endothelial changes that would identify persons likely to develop acute or chronic vascular injury are needed. Markers of endothelial dysfunction that demonstrate toxicant specificity should be sought. More information is needed on how endothelial barrier function is correlated with nonbarrier functions. Refinements in techniques are needed to render them applicable to the screening of large numbers of people for vascular function.

5. Immunologic, biochemical, cytologic, and structural markers identified as related to specific lung injury need to be correlated with physiologic measures of respiratory function, airway reactivity, particle clearance, and indexes of air-blood barriers. Understanding of those relationships could be important in developing methods for assessing risks associated with environmental exposures.

STRUCTURE AND FUNCTION

1. Animal studies are needed for increasing understanding of the pathogenic sequelae related to particle dose at specific sites in the lung. Examples of some analytic methods that can be made highly site-specific and cell-specific are morphometry, immunocytochemistry, histochemistry, and in situ hybridization.

2. Research is required on the specific cell kinetics of response to injury. Labeling indexes determined by autoradiography are not adequate for describing cell kinetics. New techniques, such as a combination of autoradiography with morphometry to measure cell pool sizes before and after injury, can make it possible to determine changes in the entire cell cycle during lung injury.

3. Three-dimensional reconstruction of cells and tissues could establish changes in intracellular organelles and cell-cell relationships that result from exposure and injury. Such techniques as computer-assisted tissue reconstructions, time-lapse photography, and high-voltage electron microscopy can be applied to obtain data on cell function and cell regulation.

4. Cell and organ culture models should be developed for extrapolating animal data on histologic changes to humans. Findings on early histologic markers of exposure and injury in animals are difficult to apply to humans, because they require invasive techniques. New ways to maintain and use human cells obtained by BAL, transbronchial lung biopsy, and tracheal explanation need to be developed.

CELLULAR AND BIOCHEMICAL RESPONSE

1. Bronchoalveolar lavage (BAL) has proved useful for evaluating lung inflammation, but further research is required

to determine its utility for assessing pollutant exposure. Interspecies studies are needed to determine relationships between changes in BAL constituents and site-specific and pollutant-specific injury. Where applicable, clinical studies should be used for confirmation of results.

2. Cell and mediator changes found in BAL and nasal-wash fluid need to be related to physiologic and pathologic changes to assess their utility as biologic markers. Furthermore, lavage fluids should be analyzed to determine whether exposure to particles or gases changes chemotactic activity. Alterations could be due to depletion or activation of pulmonary C5a, oxidants, arachidonic acid metabolites, growth factors, and other chemotactic factors that might be important markers of response.

3. Nasal lavage needs to be investigated as a means of evaluating pollutant exposure and as an epidemiologic tool. The characteristics of BAL-fluid, nasal-lavage fluid, and whole lung specimens need to be correlated in humans and animal models.

4. Monoclonal antibodies and molecular genetic techniques need to be applied to characterize types and functions of cells of the respiratory tract. As those techniques are introduced, relationships between phenotypic changes, pollutant exposure, and cell function should be established.

5. It would be of value to identify markers of susceptibility. Changes in cells and mediators in lavage fluid should be examined as possible markers of susceptibility.

6. Early markers of late-stage disease should be developed to serve as molecular probes of mechanisms of health impairment.

References

Abraham, J. L. 1978. Recent advances in pneumoconiosis: The pathologists role in etiologic diagnosis, pp. 96–137. In W. M. Thurlbeck and M. R. Abell, Eds. Lung: Structure, Function, and Disease. Baltimore: Williams and Wilkins.

Adamson, I. Y., and D. H. Bowden. 1986. Crocidolite-induced pulmonary fibrosis in mice. Cytokinetic and biochemical studies. Am. J. Pathol. 122:261–267.

Adriaenssens, P. I., C. M. White, and M. W. Anderson. 1983. Dose-response relationships for the binding of benzo(a)pyrene metabolites to DNA and protein in lung, liver, and forestomach of control and butylated hydroxyanisole-treated mice. Cancer Res. 43:3712–3719.

Agnew, J. E. 1984. Aerosol contributions to the investigation of lung structure and ventilatory function, pp. 92–126. In S. W. Clarke and D. Pavia, Eds. Aerosols and the Lung: Clinical and Experimental Aspects. London: Butterworths.

Agnew, J. E., D. Pavia, and S. W. Clarke. 1981. Airways penetration of inhaled radioaerosol: An index to small airways function? Eur. J. Respir. Dis. 62:239–255.

Agnew, J. E., P. P. Sutton, D. Pavia, and S. W. Clarke. 1986. Radioaerosol assessment of mucociliary clearance: Towards definition of a normal range. Br. J. Radiol. 59:147–151.

Ahmed, I. H., E. El-Khatib, J. W. Logus, G. C. W. Man, J. Jacques, and S. F. P. Man. 1986. Altered pulmonary epithelial permeability in canine radiation lung injury. J. Appl. Physiol. 61:971–981.

Albert, R. E., and L. C. Arnett. 1955. Clearance of radioactive dust from the human lung. AMA Arch. Ind. Health 12:99–106.

Albert, R. E., J. R. Spiegelman, M. Lippmann, and R. Bennett. 1968. The characteristics of bronchial clearance in the miniature donkey. Arch. Environ. Health 17:50–58.

Albert, R. E., M. Lippmann, and W. Briscoe. 1969. The characteristics of bronchial clearance in humans and the effects of cigarette smoking. Arch. Environ. Health 18:738–755.

Albert, R. E., M. Lippmann, H. T. Peterson, Jr., J. Berger, K. Sanborn, and D. E. Bohning. 1973. Bronchial deposition and clearance of aerosols. Arch. Intern. Med. 131:115–127.

Albert, R. E., J. Berger, K. Sanborn, and M. Lippmann. 1974. Effects of cigarette smoke components on bronchial clearance in the donkey. Arch. Environ. Health 29:96–101.

Albertini, R. J. 1980. Drug-resistant lymphocytes in man as indicators of somatic cell mutation. Teratogenesis Carcinog. Mutagen. 1:25–48.

Alpert, S. M., B. B. Schwartz, S. D. Lee, and T. R. Lewis. 1971. Alveolar protein accumulation. Arch. Intern. Med. 128:69–73.

Altshuler, B., E. D. Palmes, L. Yarmus, and N. Nelson. 1959. Intrapulmonary mixing of gases studied with aerosols. J. Appl. Physiol. 14:321–327.

American Thoracic Society. 1986. Evaluation of impairment/disability secondary to respiratory disorders. Am. Rev. Respir. Dis. 133:1205–1209.

American Thoracic Society. 1987. Single-breath carbon monoxide diffusing capacity (transfer factor): Recommendations for a standard technique. Am. Rev. Respir. Dis. 136:1299–1307.

Amoruso, M. A., G. Witz, and B. D. Goldstein. 1981. Decreased superoxide anion radical production by rat alveolar macrophages following inhalation of ozone or nitrogen dioxide. Life Sci. 28:2215–2221.

Amrein, R., R. Keller, H. Joos, and H. Herzog. 1970. New theoretical values in exploration of the respiratory function of the lung. Bull. Physiopathol. Respir. (Nancy) 6:317–349.

Andersen, M. E. 1981. A physiologically based toxicokinetic description of the metabolism of inhaled gases and vapors: Analysis at steady state. Toxicol. Appl. Pharmacol. 60:509–526.

Andersen, M. E., and J. C. Ramsey. 1983. A physiologically-based description of the inhalation pharmacokinetics of styrene in rats and humans. Dev. Toxicol. Environ. Sci. 11:415–418.

Andersen, M. E., H. J. Clewell, M. L. Gargas, F. A. Smith, and R. H. Reitz. 1987. Physiologically based pharmacokinetics and the risk assessment process for methylene chloride. Toxicol. Appl. Pharmacol. 87:185–205.

Anderson, H. A. 1985. Utilization of adipose tissue biopsy in characterizing human halogenated hydrocar-

137

bon exposure. Environ. Health Perspect. 60:127–131.

Anthonisen, N. R., J. Danson, P. C. Robertson, and W. R. D. Ross. 1969. Airway closure as a function of age. Respir. Physiol. 8:58–65.

Appleton, B. S., M. P. Goetchius, and T. C. Campbell. 1982. Linear dose-response curve for the hepatic macromolecular binding of aflatoxin B_1 in rats at very low exposures. Cancer Res. 42:3659–3662.

Ashby, J. 1982. Screening chemicals for mutagenicity: Practices and pitfalls, pp. 1–33. In J. A. Heddle, Ed. Mutagenicity: New Horizons in Genetic Toxicology. New York: Academic Press.

Au, W. W., P. Bibbins, J. B. Ward, Jr., and M. S. Legator. 1988. Development of a rodent lung macrophage chromosome aberration assay. Mutat. Res. 208:1–7.

Aubas, P., B. Cosso, P. Godard, F. B. Michel, and J. Clot. 1984. Decreased suppressor cell activity of alveolar macrophages in bronchial asthma. Am. Rev. Respir. Dis. 130:875–878.

Ayers, L. N., M. L. Ginsberg, J. Fein, and K. Wasserman. 1975. Diffusing capacity, specific diffusing capacity and interpretation of diffusion defects. West. J. Med. 123:255–264.

Babior, B. M. 1978. Oxygen-dependent microbial killing by phagocytes. N. Engl. J. Med. 298:659–668.

Bailey, M. R., F. A. Fry, and A. C. James. 1982. The long-term clearance kinetics of insoluble particles from the human lung. Ann. Occup. Hyg. 26:273–290.

Bailey, M. R., A. Hodgson, and H. Smith. 1985. Respiratory tract retention of relatively insoluble particles in rodents. J. Aerosol Sci. 16:279–293.

Baker, R. M., D. M. Burnette, R. Mankovitz, L. H. Thompson, G. F. Whitmore, L. Siminovitch, and M. J. Till. 1974. Ouabain-resistant mutants of mouse and hamster cells in culture. Cell 1:9–21.

Bang, B. G., A. L. Mukherjee, and F. B. Bang. 1967. Human nasal mucus flow rates. Johns Hopkins Med. J. 121:38–48.

Bang, F. B., and M. Foard. 1964. Interaction of respiratory epithelium of the chick and Newcastle disease virus. Am. J. Hyg. 79:260–278.

Bargon, J., H. Kronenberger, L. Bergmann, R. Buhl, J. Meier-Sydow, and P. Mitrou. 1986. Lymphocyte transformation test in a group of foundry workers exposed to beryllium and non-exposed controls. Eur. J. Respir. Dis. 146(Suppl.):211–215.

Barnes, P. J., and M. J. Brown. 1981. Venous plasma histamine in exercise- and hyperventilation-induced asthma in man. Clin. Sci. 61:159–162.

Barrett, A. J. 1977a. Cathepsin B and other proteinases, pp. 181–208. In A. J. Barrett, Ed. Proteinases in Mammalian Cells and Tissues. Research Monographs in Cell and Tissue Physiology, Vol. 2. Amsterdam: North-Holland.

Barrett, A. J. 1977b. Cathepsin D and other proteinases, pp. 209–248. In A. J. Barrett, Ed. Proteinases in Mammalian Cells and Tissues. Research Monographs in Cell and Tissue Physiology, Vol. 2. Amsterdam: North-Holland.

Barrow, C. S., W. H. Steinhagen, and J. C. F. Chang. 1983. Formaldehyde sensory irritation, pp. 16–25. In J. E. Gibson, Ed. Formaldehyde Toxicity. Washington, D.C.: Hemisphere.

Barry, B. E., and J. D. Crapo. 1985. Applications of morphometric methods to study diffuse and focal injury in the lung caused by toxic agents. CRC Crit. Rev. Toxicol. 14:1–32.

Barter, C. E., and A. H. Campbell. 1976. Relationship of constitutional factors and cigarette smoking to decrease in 1-second forced expiratory volume. Am. Rev. Respir. Dis. 113:305–314.

Bascom, R., M. Wachs, R. M. Naclerio, U. Pipkorn, S. J. Galli, and L. M. Lichtenstein. 1988. Basophil influx occurs after nasal antigen challenge: Effects of topical corticosteroid pretreatment. J. Allergy Clin. Immunol. 81:580–589.

Bates, D. V. 1989. Respiratory Function in Disease, 3rd ed. Philadelphia: W. B. Saunders. 558 pp.

Baudendistel, L. J., J. B. Shields, and D. L. Kiminski. 1982. Comparisons of double indicator thermodilution measurements of extravascular lung water (EVLW) with radiographic estimation of lung water in trauma patients. J. Trauma 22:983–988.

Bauer, M. A., M. J. Utell, P. E. Morrow, D. M. Speers, and F. R. Gibb. 1986. Inhalation of 0.30 ppm nitrogen dioxide potentiates exercise-induced bronchospasm in asthmatics. Am. Rev. Respir. Dis. 134:1203–1208.

Bauer, W., M. K. Gorny, H. R. Baumann, and A. Morell. 1985. T-lymphocyte subsets and immunoglobulin concentrations in bronchoalveolar lavage of patients with sarcoidosis and high and low intensity alveolitis. Am. Rev. Respir. Dis. 132:1060–1065.

Baughman, R. P., C. H. Bosken, R. G. Loudon, P. Hurtubise, and T. Wesseler. 1983. Quantitation of bronchoalveolar lavage with methylene blue. Am. Rev. Respir. Dis. 128:266–270.

Baughman, R. P., S. Strohofer, and C. K. Kim. 1986a. Variation of differential cell counts of bronchoalveolar lavage fluid. Arch. Pathol. Lab. Med. 110:341–343.

Baughman, R. P., B. C. Corser, S. Strohofer, and D. Hendricks. 1986b. Spontaneous hydrogen peroxide release from alveolar macrophages of some cigarette smokers. J. Lab. Clin. Med. 107:233–237.

Baumgarten, C. R., A. G. Togias, R. M. Naclerio, L. M. Lichtenstein, P. S. Norman, and D. Proud. 1985. Influx of kininogens into nasal secretions after antigen challenge of allergic individuals. J. Clin. Invest. 76:191–197.

Beck, B. D., J. D. Brain, and D. E. Bohannon. 1981. The pulmonary toxicity of an ash sample from the Mt. St. Helens volcano. Exp. Lung Res. 2:289–301.

Beck, B. D., J. D. Brain, and D. E. Bohannon. 1982. An in vivo hamster bioassay to assess the toxicity of particles for the lungs. Toxicol. Appl. Pharmacol. 66:9–29.

Beck, B. D., H. A. Feldman, J. D. Brain, T. J. Smith, M. Hallock, and B. Gerson. 1987. The pulmonary toxicity of talc and granite dust as estimated from an *in vivo* hamster bioassay. Toxicol. Appl. Pharmacol. 87:222–234.

Beer, D. J., S. Matloff, and R. E. Rocklin. 1984. The influence of histamine on immune and inflammatory responses. Adv. Immunol. 35:209–268.

Begin, R., M. Rola-Pleszczynski, S. Masse, D. Nadeau, and G. Drapeau. 1983. Assessment of progression of asbestosis in the sheep model by bronchoalveolar lavage and pulmonary function tests. Thorax 38:449–457.

Belinsky, S. A., and M. W. Anderson. 1987. Tissue and cellular specificity for DNA adduct formation and persistance following *in vivo* exposure to chemicals. Prog. Exp. Tumor Res. 31:11–20.

Bell, D. Y., and G. E. R. Hook. 1979. Pulmonary alveolar proteinosis: Analysis of airway and alveolar proteins. Am. Rev. Respir. Dis. 119:979–990.

Bell, D. Y., J. A. Haseman, A. Spock, G. McLennan, and G. E. R. Hook. 1981. Plasma proteins of the bronchoalveolar surface of the lungs of smokers and nonsmokers. Am. Rev. Respir. Dis. 124:72–79.

Bellanti, J. A., Ed. 1985. Immunology III. Philadelphia: Saunders. 598 pp.

Benson, J. M., R. F. Henderson, R. O. McClellan, R. L. Hanson, and A. H. Rebar. 1986. Comparative acute toxicity of four nickel compounds to F344 rat lung. Fundam. Appl. Toxicol. 7:340–347.

Berend, N. 1982. The correlation of lung structure with function. Lung 160:115–130.

Berend, N., J. L. Wright, W. M. Thurlbeck, G. E. Marlin, and A. J. Woolcock. 1981a. Small airways disease: Reproducibility of measurements and correlation with lung function. Chest 79:263–268.

Berend, N., C. Skoog, and W. M. Thurlbeck. 1981b. Single-breath nitrogen test in excised human lungs. J. Appl. Physiol. 51:1568–1573.

Bevilacqua, M. P., J. S. Prober, M. E. Wheeler, R. S. Cotran, and M. A. Gimbrone, Jr. 1985. Interleukin-1 acts on cultured human vascular endothelium to increase the adhesion of polymorphonuclear leukocytes, monocytes, and related leukocyte cell lines. J. Clin. Invest. 76:2003–2011.

Bhalla, D. K., and T. T. Crocker. 1986. Tracheal permeability in rats exposed to ozone: An electron microscopic and autoradiographic analysis of the transport pathway. Am. Rev. Resp. Dis. 134:572–579.

Bhalla, D. K., R. C. Mannix, M. T. Kleinman, and T. T. Crocker. 1986. Relative permeability of nasal, tracheal, and bronchoalveolar mucosa to macromolecules in rats exposed to ozone. J. Toxicol. Environ. Health 17:269–283.

Bice, D. E. 1985. Methods and approaches for assessing immunotoxocity of the lower respiratory tract, pp. 145–157. In J. Dean, M. I. Luster, A. E. Munson, and H. Amos, Eds. Immunotoxicology and Immunopharmacology. Target Organ Toxicology Series. New York: Raven Press.

Bice, D. E., and B. A. Muggenburg. 1986. Localized stimulation of memory cells in secondary lung immune responses. Am. Rev. Respir. Dis. 133(4 Pt. 2):A97. (Abstract.)

Bice, D. E., D. L. Harris, J. O. Hill, B. A. Muggenburg, and R. K. Wolff. 1980a. Immune responses after localized lung immunization in the dog. Am. Rev. Respir. Dis. 122:755–760.

Bice, D. E., D. L. Harris, and B. A. Muggenburg. 1980b. Regional immunologic responses following localized deposition of antigen in the lung. Exp. Lung Res. 1:33–41.

Bice, D. E., M. A. Degen, D. L. Harris, and B. A. Muggenburg. 1982. Recruitment of antibody-forming cells in the lungs after local immunization is nonspecific. Am. Rev. Respir. Dis. 126:635–639.

Bignon, J., P. Chahinian, G. Feldmann, and C. Sapin. 1975. Ultrastructural immunoperoxidase demonstration of autologous albumin in the alveolar capillary membrane and in the alveolar lining material in normal rats. J. Cell. Biol. 64:503–509.

Bignon, J., M. C. Jaurand, P. Sébastien, and G. Dufour. 1983. Interactions of pleural tissue and cells with mineral fibers, pp. 198–207. In J. Chrétien and A. Hirsch, Eds. Diseases of the Pleura. New York: Masson.

Bitterman, P. B., S. Adelberg, and R. G. Crystal. 1983. Mechanisms of pulmonary fibrosis. Spontaneous release of the alveolar macrophage-derived growth factor in the interstitial lung disorders. J. Clin. Invest. 72:1801–1813.

Bitterman, P. B., M. D. Wewers, S. I. Rennard, S. Adelberg, and R. G. Crystal. 1986. Modulation of alveolar macrophage-driven fibroblast proliferation by alternative macrophage mediators. J. Clin. Invest. 77:700–708.

Block, E. R., and A. B. Fisher. 1977. Depression of serotonin clearance by rat lungs during oxygen exposure. J. Appl. Physiol. 42:33–38.

Bloom, A. D., and N. W. Paul, Eds. 1981. Guidelines for Studies of Human Populations Exposed to Mutagenic and Reproductive Hazards. Proceedings of a conference held January 26–27 in Washington, D.C. White Plains, N.Y.: March of Dimes Birth Defects Foundation.

Bohning, D. E., H. L. Atkins, and S. H. Cohn. 1982. Long-term particle clearance in man: Normal and impaired. Ann. Occup. Hyg. 26:259–271.

Bond, J. A., M. M. Butler, M. A. Medinsky, B. A. Muggenburg, and R. O. McClellan. 1984. Dog pulmonary macrophage metabolism of free and particle-associated (^{14}C) benzo[a]pyrene. J. Toxicol. Environ. Health 14:181–189.

Bond, J. A., A. R. Dahl, R. F. Henderson, J. S. Dutcher, J. L. Mauderly, and L. S. Birnbaum. 1986. Species differences in the disposition of inhaled butadiene. Toxicol. Appl. Pharmacol. 84:617–627.

Bond, J. A., R. K. Wolff, J. R. Harkema, J. L. Mauderly, R. F. Henderson, W. C. Griffith, and R. O. McClellan. 1988. Distribution of DNA adducts in the respiratory tract of rats exposed to diesel exhaust. Toxicol. Appl.

Pharmacol. 96:336—346.

Booij-Noord, H., N. G. M. Orie, and K. De Vries. 1971. Immediate and late bronchial obstructive reactions to inhalation of house dust and protective effects of disodium cromoglycate and prednisolone. J. Allergy Clin. Immunol. 48:344—354.

Booker, D. V., A. C. Chamberlain, J. Rundo, D. C. F. Muir, and M. L. Thomson. 1967. Elimination of 5-mu particles from the human lung. Nature 215(96):30—33.

Boucher, R. C. 1980. Chemical modulation of airway epithelial permeability. Environ. Health Perspect. 35:3—12.

Boucher, R. C. 1981. Mechanisms of pollutant-induced airways toxicity. Clin. Chest Med. 2:377—392.

Boucher, R. C., V. Ranga, P. D. Paré, L. A. Moroz, and J. C. Hogg. 1977a. The effect of allergic bronchoconstriction on respiratory mucosal permeability. Physiologist 20(4):11. (Abstract.)

Boucher, R. C., P. D. Paré, N. J. Gilmore, L. A. Moroz, and J. C. Hogg. 1977b. Airway mucosal permeability in the Ascaris suum-sensitive rhesus monkey. J. Allergy Clin. Immunol. 60:134—140.

Boucher, R. C., B. Ranga, P. D. Paré, S. Inoue, L. A. Moroz, and J. C. Hogg. 1978. Effect of histamine and methacholine on guinea pig tracheal permeability to HRP. J. Appl. Physiol. 45:939—948.

Boucher, R. C., P. D. Paré, and J. C. Hogg. 1979. Relationship between airway hyperactivity and hyperpermeability in ascaris sensitive monkeys. J. Allergy Clin. Immunol. 64:197—201.

Boucher, R. C., J. Johnson, S. Inoue, W. Hulbert, and J. C. Hogg. 1980. The effect of cigarette smoke on the permeability of guinea pig airways. Lab. Invest. 43:94—100.

Boushey, H. A., M. J. Holtzman, J. R. Sheller, and J. A. Nadel. 1980. Bronchial hyperreactivity. Am. Rev. Respir. Dis. 121:389—413.

Bowden, D. H. 1983. Cell turnover in the lung. Am. Rev. Respir. Dis. 128(2 Pt. 2):S46—S48.

Bradley, M. O., G. Dysart, K. Fitzsimmons, P. Harbach, J. Lewin, and G. Wolf. 1982. Measurement by filter elution of DNA single- and double-strand breaks in rat hepatocytes: Effect of nitrosamines and gamma-irradiation. Cancer Res. 42:2592—2597.

Brain, J. D. 1980. Macrophage damage in relation to the pathogenesis of lung diseases. Environ. Health Perspect. 35:21—28.

Brain, J. D., and R. Frank. 1973. Alveolar macrophage adhesion: Wash electrolytic composition and free cell yield. J. Appl. Physiol. 34:75—80.

Brain, J. D., and G. A. Mensah. 1983. Comparative toxicology of the respiratory tract. Am. Rev. Respir. Dis. 128(2 Pt. 2):S87—S90.

Brandtzaeg, P. 1984. Immune functions of human nasal mucosa and tonsils in health and disease, pp. 28—95. In J. Bienenstock, Ed. Immunology of the Lung and Upper Respiratory Tract. New York: McGraw-Hill.

Brennand, J., A. C. Chinault, D. S. Konecki, D. W. Mel-

ton, and C. T. Caskey. 1982. Cloned cDNA sequences of the hypoxanthine/guanine phosphoribosyl transferase gene from a mouse neuroblastoma cell line found to have amplified genomic sequences. Proc. Natl. Acad. Sci. USA 79:1950—1954.

Britt, E. J., B. Cohen, H. Menkes, E. Bleecker, S. Permutt, R. Rosenthal, and P. Norman. 1980. Airways reactivity and functional deterioration in relatives of COPD patients. Chest 77:260—261.

Brody, A. R. 1984. Inhaled particles in human disease and animal models: Use of electron beam instrumentation. Environ. Health Perspect. 56:149—162.

Brody, A. R., and L. H. Overby. 1989. Incorporation of tritiated thymidine by epithelial and interstitial cells in bronchiolar-alveolar regions of asbestos-exposed rats. Am. J. Pathol. 134:133—140.

Brody, A. R., and M. W. Roe. 1983. Deposition pattern of inorganic particles at the alveolar level in the lungs of rats and mice. Am. Rev. Respir. Dis. 128:724—729.

Brody, A. R., L. H. Hill, B. Adkins, Jr., and R. W. O'Conner. 1981. Chrysotile asbestos inhalation in rats: Deposition pattern and reaction of alveolar epithelium and pulmonary macrophages. Am. Rev. Respir. Dis. 123:670—679.

Brooks, S. M., M. A. Weiss, and I. L. Bernstein. 1985. Reactive airways dysfunction syndrome (RADS): Persistent asthma syndrome after high level irritant exposures. Chest 88:376—384.

Buist, A. S., and B. B. Ross. 1973. Quantitative analysis of the alveolar plateau in the diagnosis of early airway obstruction. Am. Rev. Respir. Dis. 108:1078—1087.

Butler, W. T., T. A. Waldmann, and R. D. Rossen. 1970. Changes in IgA and IgG concentrations in nasal secretions prior to appearance of antibody during viral respiratory infection in man. J. Immunol. 105:584—591.

Calleman, C. J., L. Ehrenberg, B. Jansson, S. Osterman-Golkar, D. Segerback, K. Svensson, and C. A. Wachmeister. 1978. Monitoring for risk assessment by means of alkyl groups in hemoglobin in persons occupationally exposed to ethylene oxide. J. Pathol. Toxicol. 2:427—442.

Camner, P., and K. Philipson. 1971. Intra-individual studies of tracheobronchial clearance in man using fluorocarbon resin particles tagged with ^{18}F and ^{99m}Tc, pp. 157—163. In W. H. Walton, Ed. Inhaled Particles III, Vol. 1. Surrey: Unwin.

Camner, P., and K. Philipson. 1978. Human alveolar deposition of 4-μ Teflon particles. Arch. Environ. Health 33:181—185.

Camner, P., K. Philipson, L. Friberg, B. Holma, B. Larsson, and J. Svedberg. 1971. Human tracheobronchial clearance studies with fluorocarbon resin particles tagged with ^{18}F. Arch. Environ. Health 22:444—449.

Camus, A. M., A. Aitio, N. Sabadie, J. Wahrendorf, and H. Bartsch. 1984. Metabolism and urinary excretion of mutagenic metabolites of benzo[a]pyrene in C57 and DBA mice strains. Carcinogenesis 5:35—39.

Cantin, A., and R. G. Crystal. 1985. Oxidants, antioxidants and the pathogenesis of emphysema. Eur. J.

Respir. Dis. 139(Suppl.):7–17.

Cantin, A., R. Begin, M. Rola-Pleszczynski, and R. Boileau. 1983. Heterogeneity of bronchoalveolar lavage cellularity in stage III pulmonary sarcoidosis. Chest 83:485–486.

Cantin, A., L. North, and R. G. Crystal. 1985. Glutathione is present in normal alveolar epithelial lining fluid in sufficient concentration to provide antioxidant protection to lung parenchymal cells. Am. Rev. Respir. Dis. 131(4 Pt. 2):A372. (Abstract.)

Carlock, L. R., T. D. Vo, C. R. DeHaven, and J. C. Murray. 1987. An anonymous genomic clone that detects a frequent RFLP adjacent to the D4S10 (G8) marker and Huntington's disease. Nucleic Acids Res. 12:377.

Carp, H., and A. Janoff. 1979. In vitro suppression of serum elastase-inhibitory capacity by reactive oxygen species generated by phagocytosing polymorphonuclear leukocytes. J. Clin. Invest. 63:793–797.

Center, D. M., W. Cruikshank, J. S. Berman, and D. J. Beer. 1983. Functional characteristics of histamine receptor-bearing mononuclear cells. I. Selective production of lymphocyte chemoattractant lymphokines with histamine used as a ligand. J. Immunol. 131:1854–1859.

Cerny, T., D. M. Barnes, P. Hasleton, P. V. Barber, K. Healy, W. Gullick, and N. Thatcher. 1986. Expression of epidermal growth factor receptor (EGF-R) in human lung tumors. Br. J. Cancer 54:265–269.

Ceuppens, J. L., L. M. Lacguet, G. Marien, M. Demedts, A. van den Eeckhout, and E. Stevens. 1984. Alveolar T-cells subsets in pulmonary sarcoidosis. Correlation with disease activity and effect of steroid treatment. Am. Rev. Respir. Dis. 129:563–568.

Chang, L. Y., L. H. Overby, A. R. Brody, and J. D. Crapo. 1988. Progressive lung cell reactions and extracellular matrix production after a brief exposure to asbestos. Am. J. Pathol. 131:156–170.

Chan-Yeung, M., and S. Lam. 1986. Occupational asthma. Am. Rev. Respir. Dis. 133:686–703.

Chapman, H. A., Jr., O. L. Stone, and Z. Vavrin. 1984. Degradation of fibrin and elastin by human alveolar macrophages in vitro. Characterization of a plasminogen activator and its role in matrix degradation. J. Clin. Invest. 73:806–815.

Chase, M. W. 1961. The preparation and standardization of satisfactory Kveim testing antigen. Am. Rev. Respir. Dis. 84:86–88.

Cheema, M. S., S. Groth, and C. Marriott. 1988. Binding and diffusion characteristics of ^{14}C EDTA and ^{99m}Tc DTPA in respiratory tract mucus glycoprotein from patients with chronic bronchitis. Thorax 43:669–673.

Chellman, G. J., R. D. White, R. M. Norton, and J. S. Bus. 1986. Inhibition of the acute toxicity of methyl chloride in male B6C3F1 mice by glutathione depletion. Toxicol. Appl. Pharmacol. 86:93–104.

Chen, S., V. Mahadevan, and L. Zieve. 1970a. Volatile fatty acids in the breath of patients with cirrhosis of the liver. J. Lab. Clin. Med. 75:622–627.

Chen, S., L. Zieve, and V. Mahadevan. 1970b. Mercaptans and dimethyl sulfide in the breath of patients with cirrhosis of the liver: Effect of feeding methionine. J. Lab. Clin. Med. 75:628–635.

Chopra, S. K., G. V. Taplin, D. H. Simmons, and D. Elam. 1977. Measurement of mucociliary transport velocity in the intact mucosa. Chest 71:155–158.

Chopra, S. K., G. V. Taplin, D. P. Tashkin, E. Trevor, and D. Elam. 1979. Imaging sites of airway obstruction and measuring functional responses to bronchodilator treatment in asthma. Thorax 34:493–500.

Churg, A., and J. L. Wright. 1983. Small airways disease and mineral dust exposure. Pathol. Annu. 18 (Pt. 2):233–251.

Cleare, M. J., E. G. Hughes, B. Jacoby, and J. Pepys. 1976. Immediate (type I) allergic responses to platinum compounds. Clin. Allergy 6:183–195.

Cockroft, D. W., B. A. Berscheid, and K. Y. Murdock. 1983. Unimodal distribution of bronchial responsiveness to inhaled histamine in a random human population. Chest 83:751–754.

Coffin, D. L., D. E. Gardner, R. S. Holzman, and F. J. Wolock. 1968. Influence of ozone on pulmonary cells. Arch. Environ. Health 16:633–636.

Cohen, D. 1973. Fenomagnetic contamination in the lungs and other organs of the human body. Science 180:745–748.

Cohen, D., S. F. Arai, and J. D. Brain. 1979. Smoking impairs long-term dust clearance from the lung. Science 204:514–517.

Cohen, D., T. S. Crowther, G. W. Gibbs, and M. R. Becklake. 1981. Magnetic lung measurements in relation to occupational exposure in asbestos miners and millers of Quebec. Environ. Res. 26:535–550.

Cohen, G., and A. I. Cederbaum. 1979. Chemical evidence for production of hydroxyl radicals during microsomal electron transfer. Science 204:66–68.

Cohen, M. H. 1975. Guest editorial: Lung cancer: A status report. J. Natl. Cancer Inst. 55:505–511.

Cole, F. S., W. J. Matthews, Jr., T. H. Rossing, D. J. Gash, N. A. Lichtenberg, and J. E. Pennington. 1983. Complement biosynthesis by human bronchoalveolar macrophages. Clin. Immun. Immunopathol. 27:153–159.

Cole, P., and P. Stanley. 1983. The importance of nasal symptoms to the chest physician. Eur. J. Respir. Dis. 64(Suppl. 126):145–148.

Cole, R., C. Turton, H. Lanyon, and J. Collins. 1980. Bronchoalveolar lavage for the preparation of free lung cells: Technique and complications. Br. J. Dis. Chest 74:273–278.

Coles, S. J., K. H. Neill, L. M. Reid, K. F. Austen, Y. Nii, E. J. Corey, and R. A. Lewis. 1983. Effects of leukotrienes C_4 and D_4 on glycoprotein and lysozyme secretion by human bronchial mucosa. Prostaglandins 25:155–170.

Colten, H. R., and L. P. Einstein. 1976. Complement metabolism: Cellular and humoral regulation. Transplant Rev. 32:3–11.

Colten, H. R., C. A. Alper, and F. S. Rosen. 1981. Genetics and biosynthesis of complement proteins. N. Engl. J. Med. 304:653–656.

Conkle, J. P., B. J. Camp, and B. E. Welch. 1975. Trace composition of human respiratory gas. Arch. Environ. Health 30:290–295.

Connell, J. T. 1979. A novel method to assess antihistamine and decongestant efficacy. Ann. Allergy 42:278–285.

Conner, M. W., A. E. Rogers, and M. O. Amdur. 1982. Response of guinea pig respiratory tract to inhalation of submicron zinc oxide particles generated in the presence of sulfur dioxide and water vapor. Toxicol. Appl. Pharmacol. 66:434–442.

Conner, M. W., I. Chaudhuri, A. E. Rogers, and M. O. Amdur. 1985. Kinetics of respiratory tract absorption and plasma clearance of horseradish peroxidase in guinea pigs. Fundam. Appl. Toxicol. 5:99–104.

Cormier, Y., J. Bélanger, J. Beaudoin, M. Laviolette, R. Beaudoin, and J. Herbert. 1984. Abnormal bronchoalveolar lavage in asymptomatic dairy farmers: Study of lymphocytes. Am. Rev. Respir. Dis. 130:1046–1049.

Cormier, Y., J. Bélanger, and M. Laviolette. 1986. Persistent bronchoalveolar lymphocytosis in asymptomatic farmers. Am. Rev. Respir. Dis. 133:843–847.

Cormier, Y., J. Bélanger, and M. Laviolette. 1987. Prognostic significance of bronchoalveolar lymphocytosis in farmer's lung. Am. Rev. Respir. Dis. 135:692–695.

Cosio, M., H. Ghezzo, J. C. Hogg, R. Corbin, M. Loveland, J. Dosman, and P. T. Macklem. 1978. The relations between structural changes in small airways and pulmonary-function tests. N. Engl. J. Med. 298:1277–1281.

Costabel, U., K. J. Bross, J. Marxen, and H. Matthys. 1984. T-lymphocytosis in bronchoalveolar lavage fluid of hypersensitivity pneumonitis. Changes in profile of T-cell subsets during the course of disease. Chest 85:514–518.

Cramer, W. O., and W. C. Miller. 1977. A computerized approach to closing volume determination. Ann. Allergy 38:22–26.

Crapo, R. O., and A. H. Morris. 1981. Standardized single breath normal values for carbon monoxide diffusing capacity. Am. Rev. Respir. Dis. 123:185–189.

Craven, N., G. Sidwall, P. West, D. S. McCarthy, and R. M. Cherniack. 1976. Computer analysis of the single-breath nitrogen washout curve. Am. Rev. Respir. Dis. 113:445–449.

Cresia, D. A., P. Nettesheim, and A. S. Hammons. 1973. Impairment of deep lung clearance by influenza virus infection. Arch. Environ. Health 26:197–201.

Creticos, P. S., S. P. Peters, N. F. Adkinson, Jr., R. M. Naclerio, E. C. Hayes, P. S. Norman, and L. M. Lichtenstein. 1984. Peptide leukotriene release after antigen challenge in patients sensitive to ragweed. N. Engl. J. Med. 310:1626–1630.

Crimi, D., A. Scordamaglia, P. Crimi, S. Zupo, and S. Barocci. 1983. Total and specific IgE in serum, bronchial lavage and bronchoalveoalar lavage of asthmatic patients. Allergy 38:553–559.

Crystal, R. G., W. C. Roberts, G. W. Hunninghake, J. E. Gadek, J. D. Fulmer, and B. R. Line. 1981. Pulmonary sarcoidosis: A disease characterized and perpetuated by activated lung T-lymphocytes. Ann. Intern. Med. 94:73–94.

Cuddihy, R. G., B. B. Boecker, and W. C. Griffith. 1979. Modelling the deposition and clearance of inhaled radionuclides, pp. 77–89. In International Atomic Energy Agency. Biological Implications of Radionuclides Released from Nuclear Industries: Proceedings of the International Symposium on Biological Implications of Radionuclides Released from Nuclear Industries, Vienna, 26–30 March 1979. Vol. II. Vienna: International Atomic Energy Agency.

Cumming, G., and S. J. Semple. 1980. Disorders of the Respiratory System, 2nd ed. Oxford: Blackwell Scientific Publications. 616 pp.

Cutillo, A. G., A. H. Morris, D. D. Blatter, T. A. Case, D. C. Ailion, C. H. Durney, and S. A. Johnson. 1984. Determination of lung water content and distribution by nuclear magnetic resonance. J. Appl. Physiol. 57:583–588.

Dahl, A. R., J. M. Benson, R. L. Hanson, and S. J. Rothenberg. 1984. The fractionation of environmental samples according to volatility by vacuum line-cryogenic distillation. Am. Ind. Hyg. Assoc. J. 45:193–198.

Dahl, A. R., L. S. Birnbaum, J. A. Bond, P. G. Gervasi, and R. F. Henderson. 1987. The fate of isoprene inhaled by rats: Comparison to butadiene. Toxicol. Appl. Pharmacol. 89:237–248.

Dally, M. B., J. V. Hunter, E. G. Hughes, M. Stewart, and A. J. Newman Taylor. 1980. Hypersensitivity to platinum salts: A population study. Am. Rev. Respir. Dis. 121(Suppl):230. (Abstract.)

Daniele, R. P. 1980. Immune defenses of the lung, pp. 624–632. In A. P. Fishman, Ed. Pulmonary Diseases and Disorders, Vol. 1. New York: McGraw-Hill.

Daniele, R. P., M. D. Altose, and D. T. Rowlands, Jr. 1975. Immunocompetent cells from the lower respiratory tract of normal human lungs. J. Clin. Invest. 56:986–995.

Daniele, R. P., C. H. Beacham, and D. J. Gorenberg. 1977a. The bronchoalveolar lymphocyte: Studies on the life history and lymphocyte traffic from blood to the lung. Cell. Immunol. 31:48–54.

Daniele, R. P., J. H. Dauber, M. D. Altose, D. T. Rowlands, Jr., and D. J. Gorenberg. 1977b. Lymphocyte studies in asymptomatic cigarette smokers. A comparison between lung and peripheral blood. Am. Rev. Respir. Dis. 116:997–1005.

Daniele, R. P., J. H. Dauber, and M. D. Rossman. 1980. Immunologic abnormalities in sarcoidosis. Ann. Intern. Med. 92:406–416.

Daniele, R. P., J. A. Elias, P. E. Epstein, and M. D. Rossman. 1985. Bronchoalveolar lavage: Role in the pathogenesis, diagnosis, and management of interstitial lung disease. Ann. Intern. Med. 102:93—108.

Dauber, J. H., M. D. Rossman, and R. P. Daniele. 1979. Bronchoalveolar cell populations in acute sarcoidosis: Observations in smoking and nonsmoking patients. J. Lab. Clin. Med. 94:862—871.

Davis, W. B., S. I. Rennard, P. B. Bitterman, and R. G. Crystal. 1983. Pulmonary oxygen toxicity: Early reversible changes in human alveolar structures induced by hyperoxia. N. Engl. J. Med. 309:878—883.

Davis, W. B., G. A. Fells, X. H. Sun, J. E. Gadek, A. Venet, and R. G. Crystal. 1984. Eosinophil-mediated injury to lung parenchymal cells and interstitial matrix: A possible role for eosinophils in chronic inflammatory disorders of the lower respiratory tract. J. Clin. Invest. 74:269—278.

Dawson, A. 1982. Elastic recoil and compliance, pp. 193—204. In J. L. Clausen, Ed. Pulmonary Function Testing Guidelines and Controversies: Equipment, Methods, and Normal Values. New York: Academic Press.

De Méo, M. P., G. Dumenil, A. H. Botta, M. Laget, V. Zabaloueff, and A. Mathias. 1987. Urine mutagenicity of steel workers exposed to coke oven emissions. Carcinogenesis 8:363—367.

De Monchy, J. G. R., H. F. Kauffman, P. Venge, G. H. Koeter, H. M. Jansen, H. J. Sluiter, and K. De Vries. 1985. Bronchoalveolar eosinophilia during allergen-induced late asthmatic reactions. Am. Rev. Respir. Dis. 131:373—376.

Dempsey, R. A., C. A. Dinarello, J. W. Mier, L. J. Rosenwasser, M. Allegretta, T. E. Brown, and D. R. Parkinson. 1982. The differential effects of human leukocyte pyrogen/lymphocyte-activating factor, T cell growth factor, and interferon on human natural killer activity. J. Immunol. 129:2504—2510.

DeNicola, D. B., A. H. Rebar, and R. F. Henderson. 1981. Early damage indicators in the lung. V. Biochemical and cytological response to NO_2 inhalation. Toxicol. Appl. Pharmacol. 60:301—312.

Deodhar, S. D., B. Barna, and H. S. Van Ordstrand. 1973. A study of the immunologic aspects of chronic berylliosis. Chest 63:309—313.

De Vuyst, P, P. Dumortier, E. Moulin, N. Yourassowsky, and J. C. Yernault. 1987. Diagnostic value of asbestos bodies in bronchoalveolar lavage fluid. Am. Rev. Respir. Dis. 136:1219—1224.

Diaz, P., F. R. Galleguillos, M. C. Gonzalez, C. F. A. Pantin, and A. B. Kay. 1984. Bronchoalveolar lavage in asthma: The effect of disodium cromoglycate (cromolyn) on leukocyte counts, immunoglobulins, and complement. J. Allergy Clin. Immunol. 74:41—48.

Dinarello, C. A. 1984. Interleukin-1 and the pathogenesis of the acute-phase response. N. Engl. J. Med. 311:1413—1418.

Dohn, M. N., and R. P. Baughman. 1985. Effect of changing instilled volume for bronchoalveolar lavage in patients with interstitial lung disease. Am. Rev. Respir. Dis. 132:390—392.

Dolovich, M. B. 1985. Technical factors influencing response to challenge aerosols, p. 9. In F. E. Hargreave and A. J. Woolcock, Eds. Airway Responsiveness: Measurement and Interpretation. Mississauga: Astra Pharmaceuticals Canada Ltd.

Dolovich, M. B., J. Sanchis, C. Rossman, and M. T. Newhouse. 1976. Aerosol penetrance: A sensitive index of peripheral airways obstruction. J. Appl. Physiol. 40:468—471.

Dorin, J. R., M. Novak, R. E. Hill, D. J. Brock, D. S. Secher, and V. van Heyningen. 1987. A clue to the basic defect in cystic fibrosis from cloning the CF antigen gene. Nature 326(6113):614—617.

Dozy, A. M., E. N. Forman, D. N. Abuelo, G. Barsel-Bowers, M. J. Mahoney, B. G. Forget, and Y. W. Kan. 1979. Prenatal diagnosis of homozygous alpha-thalassemia. J. Am. Med. Assoc. 241:1610—1612.

Drazen, J. M., K. F. Austen, R. A. Lewis, D. A. Clark, G. Goto, A. Marfat, and E. J. Corey. 1980. Comparative airway and vascular activities of leukotrienes C-1 and D in vivo and in vitro. Proc. Natl. Acad. Sci. USA 77:4354—4358.

Dunn, B. P. 1983. Wide-range linear dose-response curve for DNA binding of orally administered benzo[a]pyrene in mice. Cancer Res. 43:2654—2658.

Dusser, D. J., M. A. Collignon, G. Stanislas-Leguern, L. G. Barritault, J. Chretien, and G. J. Huchon. 1986a. Respiratory clearance of ^{99m}Tc-DTPA and pulmonary involvement in sarcoidosis. Am. Rev. Respir. Dis. 134:493—497.

Dusser, D. J., B. D. Minty, M. A. Collignon, D. Hinge, L. G. Barritault, and G. J. Huchon. 1986b. Regional respiratory clearance of aerosolized ^{99m}Tc-DTPA: Posture and smoking effects. J. Appl. Physiol. 60:2000—2006.

Eden, E., and G. M. Turino. 1986. Interleukin-1 secretion by human alveolar macrophages stimulated with endotoxin is augmented by recombinant immune (gamma) interferon. Am. Rev. Respir. Dis. 133:455—460.

Effros, R. M. 1985. Lung water measurements with the mean transit time approach. J. Appl. Physiol. 59:673—683.

Egan, E. A. 1980. Response of alveolar epithelial solute permeability to changes in lung inflation. J. Appl. Physiol. 49:1032—1036.

Eggleston, P. A., J. O. Hendley, and J. M. Gwaltney, Jr. 1984. Mediators of immediate hypersensitivity in nasal secretions during natural colds and rhinovirus infection. Acta Otolaryngol. 413(Suppl.):25—35.

Ehrlich, R., J. C. Findlay, J. D. Fenters, and D. E. Gardner. 1977. Health effects of short-term inhalation of nitrogen dioxide and ozone mixtures. Environ. Res. 14:223—231.

Eisenberg, P. R., J. R. Hansbrough, D. Anderson, and D. P. Schuster. 1987. A prospective study of lung water measurements during patient management in an intensive care unit. Am. Rev. Respir. Dis. 136:662—668.

El-Dorry, H. A., C. B. Pickett, J. S. MacGregor, and

R. L. Soffer. 1982. Tissue-specific expression of mRNAs for dipeptidyl carboxypeptidase isozymes. Proc. Natl. Acad. Sci. USA 79:4295–4297.

Elias, J., M. D. Rossman, R. B. Zurier, and R. P. Daniele. 1985a. Human alveolar macrophage inhibition of lung fibroblast growth. A prostaglandin-dependent process. Am. Rev. Respir. Dis. 131:94–99.

Elias, J. A., A. D. Schreiber, K. Gustilo, P. Chien, M. D. Rossman, P. J. Lammie, and R. P. Daniele. 1985b. Differential interleukin 1 elaboration by unfractionated and density fractionated human alveolar macrophages and blood monocytes: Relationship to cell maturity. J. Immunol. 135:3198–3204.

Elliott, F. J., and L. Reid. 1965. Some new facts about the pulmonary artery and its branching pattern. Clin. Radiol. 16:193.

Emmett, P. C., R. G. Love, W. J. Hannan, A. M. Millar, and C. A. Soutar. 1984. The relationship between the pulmonary distribution of inhaled fine aerosols and tests of airway function. Bull. Eur. Physiopathol. Respir. 20:325–332.

Engel, L. A. 1983. Gas mixing within the acinus of the lung. J. Appl. Physiol. 54:609–618.

Engel, L. A., and P. T. Macklem. 1977. Gas mixing and distribution in the lung, pp. 37–82. In J. G. Widdicombe, Ed. Respiratory Physiology II. International Review of Physiology, Vol. 14. Baltimore: University Park Press.

Engel, L. A., A. Grassino, and N. R. Anthonisen. 1975. Demonstration of airway closure in man. J. Appl. Physiol. 38:1117–1125.

Enterline, P. E., V. L. Henderson, and G. M. Marsh. 1987. Exposure to arsenic and respiratory cancer: A reanalysis. Am. J. Epidemiol. 125:929–938.

Epstein, P. E., J. H. Dauber, M. D. Rossman, and R. P. Daniele. 1982. Bronchoalveolar lavage in a patient with chronic berylliosis: Evidence for hypersensitivity pneumonitis. Ann. Intern. Med. 97:213–216.

Ernst, P., D. Thomas, and M. R. Becklake. 1986. Respiratory survey of North American Indian children living in proximity to an aluminum smelter. Am. Rev. Respir. Dis. 133:307–312.

Eschenbacher, W. L., and T. R. Gravelyn. 1987. A technique for isolated airway segment lavage. Chest 92:105–109.

Ettensohn, D. B., P. A. Lalor, and N. J. Roberts, Jr. 1986. Human alveolar macrophage regulation of lymphocyte proliferation. Am. Rev. Respir. Dis. 133:1091–1096.

Evans, M. J. 1982. Cell death and cell renewal in small airways and alveoli, pp. 189–218. In H. Witschi, and P. Nettesheim, Eds. Mechanisms in Respiratory Toxicology. Boca Raton, Fla.: CRC Press.

Fahey, P. J., M. J. Utell, R. J. Mayewski, J. D. Wandtke, and R. W. Hyde. 1982. Early diagnosis of bleomycin pulmonary toxicity using bronchoalveolar lavage in dogs. Am. Rev. Respir. Dis. 126:126–130.

Faith, R. E., M. I. Luster, and J. G. Vos. 1980. Effects on immunocompetence by chemicals of environmental concern, pp. 173–212. In E. Hodgson, J. R. Bend, and R. M. Philpot, Eds. Reviews in Biochemical Toxicology, Vol. 2. New York: Elsevier.

Farr, B., S. F. Hackett, B. Winther, and J. O. Hendley. 1984. A method for measuring polymorphonuclear leukocyte concentrations in nasal mucus. Acta Otolaryngol. 413(Suppl.):15–18.

Fauci, A. S., A. M. Macher, D. L. Longo, H. C. Lane, A. H. Rook, H. Masur, and E. P. Gelmann. 1984. NIH conference. Acquired immunodeficiency syndrome: Epidemiologic, clinical, immunologic and therapeutic considerations. Ann. Intern. Med. 100:92–106.

Feinberg, A. P., and B. Vogelstein. 1983. Hypomethylation distinguishes genes of some human cancers from their normal counterparts. Nature 301(5895):89–92.

Felarca, A. B., and F. C. Lowell. 1971. The accumulation of eosinophils and basophils at skin sites as related to intensity of skin reactivity and symptoms of atopic disease. J. Allergy Clin. Immunol. 48:125–133.

Felicetti, S. A., R. K. Wolff, and B. A. Muggenburg. 1981. Comparison of tracheal mucous transport in rats, guinea pigs, rabbits, and dogs. J. Appl. Physiol. 51:1612–1617.

Fels, A. O. S., and Z. A. Cohn. 1986. The alveolar macrophage. J. Appl. Physiol. 60:353–369.

Ferris, B. G. 1978. Epidemiology Standardization Project (American Thoracic Society). Am. Rev. Respir. Dis. 118(6 Pt. 2):1–120.

Ferro, T. J., J. A. Kern, J. A. Elias, M. Kamoun, R. P. Daniele, and M. D. Rossman. 1987. Alveolar macrophages, blood monocytes, and density-fractioned alveolar macrophages differ in their ability to promote lymphocyte proliferation to mitogen and antigen. Am. Rev. Respir. Dis. 135:682–687.

Finch, G. L., G. L. Fisher, T. L. Hayes, and D. W. Golde. 1982. Surface morphology and functional studies of human alveolar macrophages from cigarette smokers and nonsmokers. J. Reticuloendothel. Soc. 32:1–23.

Finley, T. N., E. W. Swenson, W. S. Curran, G. L. Huber, and A. J. Ladman. 1967. Bronchopulmonary lavage in normal subjects and patients with obstructive lung disease. Ann. Intern. Med. 66:651–658.

Fiserova-Bergerova, V., Ed. 1983. Modeling of Inhalation Exposure to Vapors: Uptake, Distribution and Elimination. Vol. I and II. Boca Raton, Fla.: CRC Press.

Fiserova-Bergerova, V., and D. A. Holaday. 1979. Uptake and clearance of inhalation anesthetics in man. Drug. Metab. Dispos. 9:43–60.

Fiserova-Bergeova, V., J. Vlach, and J. C. Cassady. 1980. Predictable "individual differences" in uptake and excretion of gases and lipid soluble vapour simulation study. Br. J. Indust. Med. 37:42–49.

Fishman, A. P., Ed. 1980. Pulmonary Diseases and Disorders. New York: McGraw-Hill. 2 vols.

Fletcher, C. M., R. Peto, C. M. Tinker, and F. E. Speizer. 1976. The Natural History of Chronic Bronchitis and Emphysema: An Eight-Year Study of Working Men in London. New York: Oxford University Press. 272 pp.

Forkert, L., S. Dhingra, and N. R. Anthonisen. 1979. Airway closure and closing volume. J. Appl. Physiol. 46:24–30.

Foster, W. M., E. G. Langenback, D. Bohning, and E. H. Bergofsky. 1978. Quantitation of mucus clearance in peripheral lung and comparison with tracheal and broncial mucus transport velocities in man: Adrenergics return depressed clearance and transport velocities in asthmatics to normal. Am. Rev. Respir. Dis. 117(Suppl.):337. (Abstract.)

Foster, W. M., E. G. Langenback, and E. H. Bergofsky. 1982. Lung mucociliary function in man: Interdependence of bronchial and tracheal mucus transport velocities with lung clearance in bronchial asthma and healthy subjects. Ann. Occup. Hyg. 26:227–244.

Foster, W. M., E. G. Langenback, and E. H. Bergofsky. 1985. Disassociation in the mucociliary function of central and peripheral airways of asymptomatic smokers. Am. Rev. Respir. Dis. 132:633–639.

Fowler, W. S. 1949. Lung function studies. III. Uneven pulmonary ventilation in normal subjects and in patients with pulmonary disease. J. Appl. Physiol. 2:283–299.

Frampton, M. W., A. M. Smeglin, N. J. Roberts, Jr., J. N. Finkelstein, P. E. Morrow, and M. J. Utell. 1987. Intermittent peak versus continuous NO_2 exposure: Effects on human alveolar macrophages. Am. Rev. Respir. Dis. 135:A58. (Abstract.)

Freedman, A. P., and S. E. Robinson. 1981. Evaluation of magnetopneumography for assessing thoracic accumulation of welding fume particulate and lung dust clearance, pp. 489–496. In S. N. Erné, H. D. Hahlbohm and H. Lübbig, Eds. Biomagnetism. Proceedings Third International Workshop on Biomagnetism, Berlin (West), May 1980. Berlin: de Gruyter.

Freedman, A. P., S. E. Robinson, and R. J. Johnston. 1980. Non-invasive magnetopneumographic estimation of lung dust loads and distribution in bituminous coal workers. J. Occup. Med. 22:613–618.

Freedman, A. P., S. E. Robinson, and F. H. Y. Green. 1982. Magnetopneumography as a tool for the study of dust retention in the lungs. Ann. Occup. Hyg. 26:319–335.

Freeman, B. A., and J. D. Crapo. 1981. Hyperoxia increases oxygen radical production in rat lungs and lung mitochondria. J. Biol. Chem. 256:10986–10992.

Freeman, B. A., and J. D. Crapo. 1982. Free radicals and tissue injury. Lab Invest. 47:412–426.

Friedman, M., F. D. Stott, D. O. Poole, R. Dougherty, G. A. Chapman, H. Watson, and M. A. Sackner. 1977. A new roentenographic method for estimating mucous velocity in airways. Am. Rev. Respir. Dis. 115:67–72.

Friend, S. H., R. Bernards, S. Rogelj, R. A. Weinberg, J. M. Rapaport, D. M. Albert, and T. P. Dryja. 1986. A human DNA segment with properties of the gene that predisposes to retinoblastoma and osteosarcoma.

Nature 323(6089):643–646.

Gadek, J. E., G. A. Fells, D. G. Wright, and R. G. Crystal. 1980. Human neutrophil elastase functions as a type III collagen "collagenase." Biochem. Biophys. Res. Commun. 95:1815–1822.

Gadek, J. E., G. A. Fells, R. L. Zimmerman, S. I. Rennard, and R. G. Crystal. 1981. Antielastases of the human alveolar structures. Implications for the protease-antiprotease theory of emphysema. J. Clin. Invest. 68:889–898.

Galdston, M., V. Levytska, V., M. S. Schwartz, and B. Magnusson. 1984. Ceruloplasmin: Increased serum concentration and impaired antioxidant activity in cigarette smokers, and ability to prevent suppression of elastase inhibitory capacity of alpha 1-proteinase inhibitor. Am. Rev. Respir. Dis. 129:258–263.

Gamou, S., J. Hunts, H. Harigai, S. Hirohashi, Y. Shimosato, I. Pastan, and N. Shimizu. 1987. Molecular evidence for the lack of epidermal growth factor receptor gene expression in small cell lung carcinoma cells. Cancer Res. 47:2668–2673.

Gamsu, G., R. M. Weintraub, and J. A. Nadel. 1973. Clearance of tantalum from airways of different caliber in man evaluated by a roentgenographic method. Am. Rev. Respir. Dis. 107:214–224.

Ganey, P. E., and R. A. Roth. 1987. Elevated serum copper concentration in monocrotaline pyrrole treated rats with pulmonary hypertension. Biochem. Pharmacol. 36:3535–3537.

Garcia, J. G. N., R. G. Wolven, P. L. Garcia, and B. A. Keogh. 1986. Assessment of interlobar variation of bronchoalveolar lavage cellular differentials in interstitial lung diseases. Am. Rev. Respir. Dis. 133:444–449.

Gearhart, J. M., and R. B. Schlesinger. 1987. A morphometric analysis of rabbit airways after repeated exposure to sulfuric acid aerosol. Toxicologist 7:193.

Gearhart, J. M., and R. B. Schlesinger. 1988. Response of the tracheobronchial mucociliary clearance system to repeated irritant exposure: Effect of sulfuric acid mist on function and structure. Exp. Lung Res. 14:587–606.

Gebhart, J., J. Heyder, and W. Stahlhofen. 1981. Use of aerosols to estimate pulmonary air-space dimensions. J. Appl. Physiol. 51:465–476.

Geggel, R. L., A. C. A. Carvalho, L. W. Hoyer, and L. M. Reid. 1987. Von Willebrand factor abnormalities in primary pulmonary hypertension. Am. Rev. Respir. Dis. 135:294–299.

Gelb, A. F., and N. Zamel. 1973. Simplified diagnosis of small airway obstruction. N. Engl. J. Med. 288:395–398.

Gerberick, G. H., H. A. Jaffe, J. B. Willoughby, and W. F. Willoughby. 1986. Relationships between pulmonary inflammation, plasma transudation, and oxygen metabolite secretion by alveolar macrophages. J. Immunol. 137:114–121.

Gillis, C. N., and B. R. Pitt. 1982. The fate of circulating amines within the pulmonary circulation. Annu. Rev.

Physiol. 44:269—281.

Ginns, L. C., P. D. Goldenheim, R. C. Burton, R. B. Colvin, L. G. Miller, G. Goldstein, C. Hurwitz, and H. Kazemi. 1982. T-lymphocyte subsets in peripheral blood and lung lavage in idiopathic pulmonary fibrosis and sarcoidosis: Analysis by monoclonal antibodies and flow cytometry. Clin. Immunol. Immunopathol. 25:11—20.

Gleich, G. J., E. Frigas, D. A. Loegering, D. L. Wasson, and D. Steinmuller. 1979. Cytotoxic properties of the eosinophil major basic protein. J. Immunol. 123:2925—2927.

Godard, P., J. Clot, O. Jonquet, J. Bousquet, and F. B. Michel. 1981. Lymphocyte subpopulations in broncho-alveolar lavages of patients with sarcoidosis and hypersensitivity pneumonitis. Chest 80:447—452.

Gold, P. M. 1982. Single breath nitrogen test: Closing volume and distribution of ventilation, pp. 105—114. In J. L. Clausen, Ed. Pulmonary Function Testing Guidelines and Controversies: Equipment, Methods, and Normal Values. New York: Academic Press.

Golden, J. A., J. A. Nadel, and H. A. Boushey. 1978. Bronchial hyperirritability in healthy subjects after exposure to ozone. Am. Rev. Respir. Dis. 118:287—294.

Gongora, G., M. Roy, R. Gongora, P. Drutel, J. P. Gaillard, and H. Jammet. 1981. Bronchopulmonary clearance of radioactive labelled particles among 90 normal and pathological subjects. In H. Hauck, Ed. Proceedings of International Symposium on Deposition and Clearance of Aerosols, Bad Gleichenberg, Austria. Vienna: Arbeitsgemeinschaft fur Aerosols in der Medizine.

Gonzalez, F. J., A. K. Jaiswal, and D. W. Nebert. 1986. P450 genes: Evolution, regulation and relationship to human cancer and pharmacogenetics. Cold Spring Harbor Symp. Quant. Biol. 51(Pt. 2):879—890.

Goodman, R. M., B. M. Yergin, J. F. Landa, M. H. Golinvaux, and M. A. Sackner. 1978. Relationship of smoking history and pulmonary function tests to tracheal mucus velocity in nonsmokers, young smokers, ex-smokers and patients with chronic bronchitis. Am. Rev. Respir. Dis. 117:205—214.

Gordon, T., J. E. Thompson, and D. Sheppard. 1988. Arachidonic acid metabolites do not mediate toluene diisocyanate-induced airway hyperresponsiveness in guinea pigs. Prostaglandins 35:699—706.

Gorin, A. B., W. J. Weidner, R. H. Demling, and N. C. Staub. 1978. Noninvasive measurement of pulmonary transvascular protein flux in sheep. J. Appl. Physiol. 45:225—233.

Gosmset, P., A. B. Tonnel, M. Joseph, L. Prin, A. Mallart, J. Charon, and A. Capron. 1984. Secretion of a chemotactic factor for neutrophils and eosinophils by alveolar macrophages from asthmatic patients. J. Allergy Clin. Immunol. 74:827—834.

Graham, D., F. W. Henderson, and D. House. 1988. Neutrophil influx measured in nasal lavages of humans exposed to ozone. Arch. Environ. Health. 43:228—233.

Green, L. C., P. L. Skipper, R. J. Turesky, M. S. Bryant, and S. R. Tannenbaum. 1984. In vivo dosimetry of 4-aminobiphenyl in rats via a cysteine adduct in hemoglobin. Cancer Res. 44:4254—4259.

Greening, A. P., and D. B. Lowrie. 1983. Extracellular release of hydrogen peroxide by human alveolar macrophages. The relationship to cigarette smoking and lower respiratory tract infections. Clin. Sci. 65:661—664.

Griffin, M., J. W. Weiss, A. G. Leitch, E. R. McFadden, Jr., E. J. Corey, K. F. Austen, and J. M. Drazen. 1983. Effects of leukotriene D on the airways in asthma. N. Engl. J. Med. 308:436—439.

Griffith, W. C., R. C. Cuddihy, B. B. Boecker, R. A. Guilmette, M. A. Medinsky, and J. A. Mewhinney. 1983. Comparison of solubility of aerosols in lungs of laboratory animals. Health Phys. 45:233. (Abstract.)

Groopman, J. D., P. R. Donahue, J. Q. Zhu, J. S. Chen, and G. N. Wogan. 1985. Aflatoxin metabolism in humans: Detection of metabolites and nucleic acid adducts in urine by affinity chromatography. Proc. Natl. Acad. Sci. USA 82:6492—6496.

Groopman, J. D., B. D. Roebuck, and T. W. Kensler. 1987. Application of monoclonal antibodies and dietary antioxidant-based animal models to define human exposure to aflatoxin B_1. Prog. Exp. Tumor Res. 31:52—62.

Gross, M. L., J. O. Lay, Jr., P. A. Lyon, D. Lippstreu, N. Kansas, R. L. Harless, S. E. Taylor, and A. E. Dupuy, Jr. 1984. 2,3,7,8,-Tetrachlorodibenzo-p-dioxin levels in adipose tissue of Vietnam veterans. Environ. Res. 33:261—268.

Grover, M., R. A. Slutsky, C. B. Higgins, and R. Shabetai. 1983. Extravascular lung water in patients with congestive heart failure. Radiology 147:659—662.

Gupta, R. C. 1987. [32]p-Postlabeling assay to measure carcinogen-DNA adducts. Prog. Exp. Tumor Res. 31:21—32.

Hadfield, E. H., and R. G. MacBeth. 1971. Adenocarcinoma of ethmoids in furniture workers. Ann. Otol. Rhinol. Laryngol. 80:699—703.

Halpern, M., S. J. Williamson, D. M. Spektor, R. B. Schlesinger, and M. Lippmann. 1981. Remanent magnetic fields for measuring particle retention and distribution in the lung. Exp. Lung Res. 2:27—35.

Hance, A. J., S. Douches, R. J. Winchester, V. J. Ferrans, and R. G. Crystal. 1985. Characterization of mononuclear phagocyte subpopulations in the human lung by using monoclonal antibodies: Changes in alveolar macrophage phenotype associated with pulmonary sarcoidosis. J. Immunol. 134:284—292.

Hanifin, J. M., W. L. Epstein, and M. J. Cline. 1970. In vitro studies of granulomatous hypersensitivity to beryllium. J. Invest. Dermatol. 55:284—288.

Hankinson, J. L., E. D. Palmes, and N. L. Lapp. 1979. Pulmonary airspace size in coal miners. Am. Rev. Respir. Dis. 119:391—397.

Hansen, J. E. 1982. Exercise testing, pp. 259—279. In J. L. Clausen, Ed. Pulmonary Function Testing Guidelines and Controversies: Equipment, Methods, and Normal Values. New York: Academic Press.

Hargreave, F. E., and A. J. Woolcock, Eds. 1985. Airway Responsiveness: Measurement and Interpretation. Mississauga: Astra Pharmaceuticals Canada.

Hargreave, F. E., J. Dolovich, and L. P. Boulet. 1983. Inhalation provocation tests. Semin. Respir. Med. 4:224–236.

Hargreave, F. E., P. M. O'Byrne, and E. H. Ramsdale. 1985. Mediators, airway responsiveness, and asthma. J. Allergy Clin. Immunol. 76(2 Pt. 2):272–276.

Harper, E. 1980. Collageneases. Annu. Rev. Biochem. 49:1063–1078.

Harris, C. C., A. Weston, J. C. Willey, G. E. Trivers, and D. L. Mann. 1987. Biochemical and molecular epidemiology of human cancer: Indicators of carcinogen exposure, DNA damage, and genetic predisposition. Environ. Health Perspect. 75:109–119.

Harris, P., and D. Heath. 1986. The Human Pulmonary Circulation: Its Form and Function in Health and Disease, 3rd ed. New York: Churchill Livingstone. 702 pp.

Hatch, G. E., H. Koren, R. Slade, K. Crissman, and J. Norwood. 1987. Effect of acute ozone (O_3) exposure on human and rat lavage fluid biochemistry. Toxicologist 7:10.

Hatton, D. V., C. S. Leach, A. E. Nicogossian, and N. di Ferrante. 1977. Collagen breakdown and nitrogen dioxide inhalation. Arch. Environ. Health 32:33–36.

Haugen, A., G. Becher, C. Benestad, K. Vahakangas, G. E. Trivers, M. J. Newman, and C. C. Harris. 1986. Determination of polycyclic aromatic hydrocarbons in the urine, benzo(a)pyrene diol epoxide-DNA adducts in lymphocyte DNA, and antibodies to the adducts in sera from coke oven workers exposed to measured amounts of polycyclic aromatic hydrocarbons in the work atmosphere. Cancer Res. 46:4178–4183.

Havill, A. M., and M. H. Gee. 1984. Role of interstitium in clearance of alveolar fluid in normal and injured lungs. J. Appl. Physiol. 57:1–6.

Heath, D., and J. E. Edwards. 1958. The pathology of hypertensive pulmonary vascular disease: A description of six grades of structural changes in the pulmonary arteries with special reference to cogenital cardiac septal defects. Circulation 18:533–547.

Hedlund, L. W., P. Vock, E. L. Effman, M. M. Lischko, and C. E. Putman. 1984. Hydrostatic pulmonary edema: An analysis of lung density changes by computed-tomography. Invest. Radiol. 19:254–262.

Hedlund, L. W., P. Vock, E. L. Effman, and C. E. Putman. 1985. Morphology of oleic acid-induced lung injury. Invest. Radiol. 20:2–8.

Heinrich, U., H. Muhle, S. Takenaka, H. Ernst, R. Fuhst, U. Mohr, F. Pott, and W. Stöber. 1986. Chronic effects on the respiratory tract of hamsters, mice and rats after long-term inhalation of high concentrations of filtered and unfiltered diesel engine emissions. J. Appl. Toxicol. 6:393–395.

Henderson, R. F. 1984. Use of bronchoalveolar lavage to detect lung damage. Environ. Health Perspect. 56:115–129.

Henderson, R. F. 1988a. Use of bronchoalveolar lavage to detect lung damage, pp. 239–268. In D. E. Gardner, J. D. Crapo, and E. J. Massaro, Eds. Toxicology of the Lung. Target Organ Toxicology Series. New York: Raven Press.

Henderson, R. F. 1988b. Bronchoalveolar lavage: A tool for assessing the health status of the lung. In R. O. McClelland and R. F. Henderson, Eds. Concepts in Inhalation Toxicology. Washington, D.C.: Hemisphere.

Henderson, R. F., A. H. Rebar, J. A. Pickrell, and G. J. Newton. 1979a. Early damage indicators in the lung. III. Biochemical and cytological response of the lung to inhaled metal salts. Toxicol. Appl. Pharmacol. 50:123–136.

Henderson, R. F., A. H. Rebar, and D. B. DeNicola. 1979b. Early damage indicators in the lungs. IV. Biochemical and cytological response of the lung to inhaled metal salts. Toxicol. Appl. Pharmacol. 51:129–135.

Henderson, R. F., J. M. Benson, F. F. Hahn, C. H. Hobbs, R. K. Jones, J. L. Mauderly, R. O. McClellan, and J. A. Pickrell. 1985a. New approaches for the evaluation of pulmonary toxicity: Bronchoalveolar lavage fluid analysis. Fundam. Appl. Toxicol. 5:451–458.

Henderson, R. F., C. H. Hobbs, F. F. Hahn, J. M. Benson, J. A. Pickrell, and S. A. Silbaugh. 1985b. A comparison of in vitro and in vivo toxicity of mineral dusts, pp. 521–527. In E. G. Beck and J. Bignon, Eds. In Vitro Effects of Mineral Dusts. Berlin: Springer-Verlag.

Henderson, R. F., J. L. Mauderly, J. A. Pickrell, F. F. Hahn, H. Muhle, and A. H. Rebar. 1987. Comparative study of bronchoalveolar lavage fluid: Effect of species. Exp. Lung Res. 13:329–342.

Henderson, R. F., J. A. Pickrell, R. K. Jones, J. D. Sun, J. M. Benson, J. L. Mauderly, and R. O. McClellan. 1988. Response of rodents to inhaled diluted diesel exhaust: Biochemical and cytological changes in bronchoalveolar lavage fluid and in lung tissue. Fundam. Appl. Toxicol. 11:546–567.

Henderson, Y., and H. W. Haggard. 1943. Noxious Gases and the Principles of Respiration Influencing their Action. New York: Reinhold. 294 pp.

Henson, P. M. 1972. Pathologic mechanisms in neutrophil-mediated injury. Am. J. Pathol. 68:593–612.

Henson, P. M., K. McCarthy, G. L. Larsen, R. O. Webster, P. C. Giclas, R. B. Dreisin, T. E. King, and J. O. Shaw. 1979. Complement fragments, alveolar macrophages, and alveolitis. Am. J. Pathol. 97:93–110.

Heyder, J. 1983. Charting human thoracic airways by aerosols. Clin. Phys. Physiol. Meas. 4:29–37.

Hislop, A., and L. Reid. 1973. Fetal and childhood development of the intrapulmonary veins in man: Branching pattern and structure. Thorax 28:313–319.

Hnatowich, D. J., W. W. Layne, and R. L. Childs. 1982. The preparation and labeling of DTPA-coupled albumin. Int. J. Appl. Radiat. Isot. 33:327–332.

Hocking, W. G., and D. W. Golde. 1979. The pulmonary

alveolar macrophage. N. Engl. J. Med. 301:580—587.

Hoffman, R. A., P. C. Kung, W. P. Hansen, and G. Goldstein. 1980. Simple and rapid measurement of human T lymphocytes and their subclasses in peripheral blood. Proc. Natl. Acad. Sci. USA 77:4914—4917.

Hogg, J. C., P. D. Paré, and R. C. Boucher. 1979. Bronchial mucosal permeability. Fed. Proc. 38:197—201.

Hoidal, J. R., and D. E. Niewoehner. 1982. Lung phagocyte recruitment and metabolic alterations induced by cigarette smoke in humans and hamsters. Am. Rev. Respir. Dis. 126:548—552.

Hoidal, J. R., J. G. White, and J. E. Repine. 1979. Influence of cationic local anesthetics on the metabolism and ultrastructure of human alveolar macrophages. J. Lab. Clin. Med. 93:857—866.

Hoidal, J. R., R. B. Fox, P. A. Lemarbe, R. Perri, and J. E. Repine. 1981. Altered oxidative metabolic responses in vitro of alveolar macrophages from asymptomatic cigarette smokers. Am. Rev. Respir. Dis. 123:85—89.

Holgate, S. T., and A. B. Kay. 1985. Mast cells, mediators and asthma. Clin. Allergy 15:221—234.

Holgate, S. T., C. Hardy, C. Robinson, R. M. Agius, and P. H. Howarth. 1986. The mast cell as a primary effector cell in the pathogenesis of asthma. J. Allergy Clin. Immunol. 77:274—282.

Holma, B. 1967a. Lung clearance of mono- and di-disperse aerosols determined by profile scanning and whole-body counting. Acta Med. Scand. 473(Suppl.):1—103.

Holma, B. 1967b. Short-term lung clearance in rabbits exposed to a radioactive bi-disperse (6 and 3 μ) polystyrene aerosol, pp. 189—201. In C. N. Davies, Ed. Inhaled Particles II. Oxford: Pergamon.

Holtzman, M. J., L. M. Fabbri, P. M. O'Byrne, B. D. Gold, H. Aizawa, E. H. Walters, S. E. Alpert, and J. A. Nadel. 1983. Importance of airway inflammation for hyperresponsiveness induced by ozone. Am. Rev. Respir. Dis. 127:686—690.

Hong, R. 1976. Immunodeficiency, pp. 620—636. In N. R. Rose and H. Friedman, Eds. Manual of Clinical Immunology. Washington, D.C.: American Society for Microbiology.

Huang, F. L., D. R. Roop, and L. M. De Luca. 1986. Vitamin A deficiency and keratin biosynthesis in cultured hamster trachea. In Vitro Cell Dev. Biol. 22:223—230.

Huff, J. E., R. L. Melnick, H. A. Solleveld, J. K. Haseman, M. Powers, and R. A. Miller. 1985. Multiple organ carcinogenicity of 1,3-butadiene in B6C3F$_1$ mice after 60 weeks of inhalation exposure. Science 227(4686):548—549.

Hughes, D. A., P. L. Halslam, P. J. Townsend, and M. Turner-Warwick. 1984. Blood and bronchoalveolar lavage T-subsets in sarcoidosis and extrinsic allergic alveolitis. Thorax 39:708—709. (Abstract.)

Hugli, T. E. 1981. The structural basis of anaphylatoxin in chemotactic functions of C3a, C4a, and C5a. CRC Crit. Rev. Immunol. 1:321—366.

Hulbert, W. C., D. C. Walker, A. Jackson, and J. C. Hogg. 1981. Airway permeability to horseradish peroxidase in guinea pigs: The repair phase after injury by cigarette smoke. Am. Rev. Respir. Dis. 123:320—326.

Hunninghake, G. W. 1984. Release of interleukin-1 by alveolar macrophages of patients with active pulmonary sarcoidosis. Am. Rev. Respir. Dis. 129:569—572.

Hunninghake, G. W., and R. G. Crystal. 1981a. Mechanisms of hypergammaglobinemina in pulmonary sarcoidosis: Site of increased antibody production and role of T lymphocytes. J. Clin. Invest. 67:86—92.

Hunninghake, G. W., and R. G. Crystal. 1981b. Pulmonary sarcoidosis: A disorder mediated by excess helper T-lymphocyte activity at sites of disease activity. N. Engl. J. Med. 305:429—434.

Hunninghake, G. W., and R. G. Crystal. 1983. Cigarette smoking and lung destruction: Accumulation of neutrophils in the lungs of cigarette smokers. Am. Rev. Respir. Dis. 128:833—838.

Hunninghake, G. W., J. D. Fulmer, R. C. Young, Jr., J. E. Gadek, and R. G. Crystal. 1979a. Localization of the immune response in sarcoidosis. Am. Rev. Respir. Dis. 120:49—57.

Hunninghake, G. W., J. E. Gadek, O. Kawanami, V. J. Ferrans, and R. G. Crystal. 1979b. Inflammatory and immune processes in the human lung in health and disease: Evaluation by bronchoalveolar lavage. Am. J. Pathol. 97:149—206.

Hunninghake, G. W., J. E. Gadek, H. M. Fales, and R. G. Crystal. 1980a. Human alveolar macrophage derived chemotactic factors for neutrophils: Stimuli and partial characterization. J. Clin. Invest. 66:473—483.

Hunninghake, G. W., B. A. Keogh, B. R. Line, J. E. Gadek, O. Kawanami, V. J. Ferrans, and R. G. Crystal. 1980b. Pulmonary sarcoidosis: Pathogenesis and therapy, pp. 275—290. In D. L. Boros and T. Yoshida, Eds. Basic and Clinical Aspects of Granulomatous Diseases. New York: Elsevier North-Holland.

Hunninghake, G. W., J. E. Gadek, R. C. Young, Jr., O. Kawanami, V. J. Ferrans, and R. G. Crystal. 1980c. Maintenance of granuloma formation in pulmonary sarcoidosis by T-lymphocytes within the lung. N. Engl. J. Med. 302:594—498.

Hunninghake, G. W., G. N. Badell, D. C. Zavala, M. Monick, and M. Brady. 1983. Role of interleukin-2 release by lung T-cells in active pulmonary sarcoidosis. Am. Rev. Respir. Dis. 128:634—638.

Hunter, J. A. A., W. E. Finkbeiner, J. A. Nadel, E. J. Goetzl, and M. J. Holtzman. 1985. Predominant generation of 15-lipoxygenase metabolites of arachidonic acid by epithelial cells from human trachea. Proc. Natl. Acad. Sci. USA 82:4633—4637.

Hurst, D. J., D. E. Gardner, and D. L. Coffin. 1970. Effect of ozone on acid hydrolases of the pulmonary alveolar macrophage. Res. J. Reticuloendothel. Soc. 8:288—301.

Hutchinson, J. 1846. On the capacity of the lungs and on the respiratory movements with a view of establishing a precise and easy method of detecting disease by the spirometer. Lancet 1:630–632.

ILO (International Labour Office). 1980. Guidelines for the Use of International Labour Office Classification of Radiographs of Pneumonioniosis. Occupational Safety and Health Sciences 22. Geneva: International Labour Office.

Iskander, M. F., C. H. Durney, and D. J. Shoff. 1979. Diagnosis of pulmonary edema by a surgically noninvasive microwave technique. Radio Sci. 14(6S):265–269.

Incalzi, R. A., R. Pistelli, V. Locci, F. Patalano, S. M. Liberatore, and G. Ciappi. 1985. Detection of ventilation uneveness by nitrogen washout and forced vital capacity manoeuvres: Its limits and reliability. Respiration 48:145–152.

Jackson, A. C., and J. W. Watson. 1982. Oscillatory mechanics of the respiratory system in normal rats. Respir. Physiol. 48:309–322.

Jackson, A. C., J. W. Watson, and M. I. Kotlikoff. 1984. Respiratory system, lung and chest wall impedances in anesthetized dogs. J. Appl. Physiol. 57:34–39.

Jacobson, G., and W. S. Lainhart. 1972. ILO U/C 1971 international classification of radiographs of pneumoconioses. Med. Radiogr. 48:67–76.

Jaiswal, A. K., F. J. Gonzalez, and D. W. Nebert. 1985. Human P1-450 gene sequence and correlation of mRNA with genetic differences in benzo[a]pyrene metabolism. Nucleic Acids Res. 13:4503–4520.

Jakab, G. J. 1984. Viral-bacterial interactions, pp. 298–310. In J. Bienenstock, Ed. Immunology of the Lung and Upper Respiratory Tract. New York: McGraw-Hill.

Janoff, A. 1972. Human granulocyte elastase. Further delineation of its role in connective tissue damage. Am. J. Pathol. 68:579–592.

Jetten, A. M., and J. E. Shirley. 1986. Characterization of transglutaminase activity in rabbit tracheal epithelial cells. Regulation by retinoids. J. Biol. Chem. 261:15097–15101.

Jetten, A. M., J. I. Rearick, and H. L. Smits. 1986. Regulation of differentiation of airway epithelial cells by retinoids. Biochem. Soc. Trans. 14:930–933.

Johnson, H. M., and W. L. Farrar. 1983. The role of a gamma interferon-like lymphokine in the activation of T cells for expression of interleukin 2 receptors. Cell Immunol. 75:154–159.

Johnson, K. J., and P. A. Ward. 1982. Acute and progressive lung injury after contact with phorbal myristate acetate. Am. J. Pathol. 107:29–35.

Johnson, K. J., J. C. Fantone, III, J. Kaplan, and P. A. Ward. 1981. In vivo damage of rat lungs by oxygen metabolites. J. Clin. Invest. 67:983–993.

Jones, J. G., B. D. Minty, P. Lawler, G. H. Hulands, J. C. Crawley, and N. Veall. 1980. Increased alveolar epithelial permeability in cigarette smokers. Lancet 1:66–68.

Jones, R., W. M. Zapol, and L. M. Reid. 1984. Pulmonary artery remodeling and pulmonary artery hyper-

tension after exposure to hyperoxia for 7 days. A morphometric and hemodynamic study. Am. J. Pathol. 117:273–285.

Joseph, M., A. B. Tonnel, G. Torpier, A. Capron, B. Arnoux, and J. Benveniste. 1983. Involvement of immunoglobulin E in the secretory processes of alveolar macrophages from asthmatic patients. J. Clin. Invest. 71:221–230.

Junttila, M. J., K. Kalliomaki, and P. L. Kalliomaki. 1983. A new magnetic method for studying the lung-retained dust in vivo. Proceedings of the VI International Pneumoconiosis Conference. September.

Kaji, H., N. M. Hisamura, N. Sato, and M. Murao. 1978. Evaluation of volatile sulfur compounds in the expired alveolar gas in patients with liver cirrhosis. Clin. Chem. Acta 85:279–284.

Kalliomaki, K., P. L. Kalliomaki, V. Kelha, and V. Vaaranen. 1980. Instrumentation for measuring the magnetic lung contamination of steel welders. Ann. Occup. Hyg. 23:175–184.

Kalliomaki, K., K. Aittoniemi, P. L. Kalliomaki, and M. Moilanen. 1981. Measurement of lung-retained continments in vivo among workers exposed to metal aerosols. Am. Ind. Hyg. Assoc. J. 42:234–238.

Kalliomaki, K., P. L. Kalliomaki, and M. Moilanen. 1983. A new magnetic method for studying the quality of lung-retained dust in vivo. Acta. Polytechnia Scand. App. Phys. (138):72–75.

Kalliomaki, K., P. L. Kalliomaki, and M. Moilanen. 1986. A mobile magnetopneumograph with dust quality sensing, pp. 215–218. In R. M. Stern, A. Berlin, A. C. Fletcher, and J. Jarvisalo, Eds. Heath Hazards and Biological Effects of Welding Fumes and Gases. International Congress Series, No. 676. Amsterdam: Excerpta Medica.

Kalliomaki, P. L., K. Alanko, O. Korhonen, T. Mattsson, V. Vaaranen, and M. Koponen. 1978a. Amount and distribution of welding fume and lung continments among arc welders. Scand. J. Work Environ. Health 4:122–130.

Kalliomaki, P. L., O. Korhonen, V. Vaaranen, K. Kalliomaki, and M. Koponen. 1978b. Lung retention and clearance of shipyard arc welders. Int. Arch. Occup. Environ. Health 42:83–90.

Kalliomaki, P. L., O. Korhonen, T. Mattsson, V. Sortti, V. Vaaranen, K. Kalliomaki, and M. Koponen. 1979. Lung contamination among foundry workers. Int. Arch. Occup. Environ. Health 43:85–91.

Kaltreider, H. B., J. L. Caldwell, and E. Adam. 1977. The fate and consequences of an organic particulate antigen instilled into bronchoalveolar spaces of normal canine lungs. Am. Rev. Respir. Dis. 116:267–280.

Kameda, K., T. Ono, and Y. Imai. 1979. Participation of superoxide, hydrogen peroxide and hydroxyl radicals in NADPH-cytochrome P-450 reductase-catalyzed peroxidation of methyl linolenate. Biochem. Biophys. Acta 572:77–82.

Kannel, W. B., E. A. Lew, H. B. Hubert, and W. P. Castelli. 1980. The value of measuring vital capacity for prognostic purposes. Trans. Assoc. Life Insur. Med. Dir. Am. 64:66–83.

Kauffmann, F. 1984. Genetics of chronic obstructive pulmonary diseases: Searching for their heterogeneity. Bull. Eur. Physiopathol. Respir. 20:163–210.

Kawano, H., T. Inamasu, M. Ishizawa, N. Ishinishi, and J. Kumazawa. 1987. Mutagenicity of urine from young male smokers and nonsmokers. Int. Arch. Occup. Environ. Health 59:1–9.

Kehrl, H. R., L. M. Vincent, R. J. Kowalsky, D. H. Horstman, J. J. O'Neil, W. H. McCartney, and P. A. Bromberg. 1987. Ozone exposure increases respiratory epithelial permeability in humans. Am. Rev. Respir. Dis. 135:1124–1128.

Kejeldgaard, J. M., R. W. Hyde, D. M. Speers, and W. W. Reichert. 1976. Frequency dependence of total respiratory resistance in early airway disease. Am. Rev. Respir. Dis. 114:501–508.

Keller, R. H., S. Swartz, D. P. Schlueter, S. Bar-Sela, and J. N. Fink. 1984. Immunoregulation in hypersensitivity pneumonitis: Phenotypic and functional studies of bronchoalveolar lavage lymphocytes. Am. Rev. Respir. Dis. 130:766–771.

Kelly, J., D. Hemenway, and J. N. Evans. 1986. Hydroxylysine as a marker of lung injury in nitrogen dioxide exposure. Am. Rev. Respir. Dis. 133:A86. (Abstract.)

Kennedy, S. M., R. K. Elwood, B. J. R. Wiggs, P. D. Paré, and J. R. Hogg. 1984. Increased airway mucosal permability of smokers: Relationship to airway reactivity. Am. Rev. Respir. Dis. 129:143–148.

Kerns, W. D., D. J. Donofrio, and K. L. Pavkov. 1983. The chronic effects of formaldehyde inhalation in rats and mice: A preliminary report, pp. 111–131. In J. E. Gibson, Ed. Formaldehyde Toxicity. Washington, D.C.: Hemisphere.

Kerr, J. S., D. J. Riley, M. M. Frank, R. L. Trelsted, and H. M. Frankel. 1984. Reduction of chronic hypoxic pulmonary hypertension in the rat by B-aminopropionitrile. J. Appl. Physiol. 57:1760–1766.

Kerr, J. S., C. L. Ruppert, C. A. Tozzi, J. A. Neubauer, H. M. Frankel, S. Y. Yu, and D. J. Riley. 1987. Reduction of chronic hypoxic pulmonary hypertension in the rat by an inhibitor of collagen production. Am. Rev. Respir. Dis. 135:300–306.

Khan, M. M., and K. Melmon. 1985. Are autocoids more than theoretic modulators of immunity? Clin. Immunol. Rev. 4:1–30.

Kidd, V. J., R. B. Wallace, K. Itakura, and S. L. Woo. 1983. Alpha 1-antitrypsin deficiency detection by direct analysis of the mutation in the gene. Nature 304(5923):230–234.

Kim, C. S., L. K. Brown, G. G. Lewars, and M. A. Sackner. 1983. Aerosol rebreathing method for assessment of airway abnormalities: Theoretical analysis and validation. Am. Ind. Hyg. Assoc. J. 44:349–357.

Kim, C. S., G. A. Lewars, and M. A. Sackner. 1985. Total aerosol deposition as an index of lung abnormality. Presented at American Association for Aerosol Research, Annual Meeting, Albuquerque, New Mexico, November 1985.

Kim, K. J., T. R. LeBon, J. S. Shinbane, and E. D. Crandall. 1985. Asymmetric ^{14}C-albumin transport across bullfrog alveolar epithelium. J. Appl. Physiol. 59:1290–1297.

Kimura, A., and E. Goldstein. 1981. Effect of ozone on concentrations of lysozyme in phagocytizing alveolar marcophages. J. Infect. Dis. 143:247–251.

Kirby, J. G., P. M. O'Byrne, and F. E. Hargreave. 1987. Bronchoalveolar lavage does not alter airway responsiveness in asthmatic subjects. Am. Rev. Respir. Dis. 135:554–556.

Klech, H., P. Haslam, and M. Turner-Warwick. 1986. World wide clinical survey on bronchoalveolar lavage (BAL) in sarcoidosis. Experience in 62 centers in 19 countries. Preliminary analysis. Sarcoidosis 3:113–117.

Knight, D. R., H. J. O'Neill, S. M. Gordon, E. H. Luebcke, and J. S. Bowman. 1984. The Body Burden of Organic Vapors in Artificial Air: Trial Measurements Aboard a Moored Submarine. Memo Report 84-4. Groton, Conn.: Naval Submarine Medical Research Laboratory.

Knight, D. R., J. O'Neill, J. S. Bowman, and S. M. Gordon. 1985. Use of the expired breath to monitor the air quality in sealed capsules. Physiologist 28:336. (Abstract.)

Knowles, M. R., J. L. Carson, A. M. Collier, J. T. Gatzy, and R. C. Boucher. 1981. Measurements of nasal transepithelial electric potential differences in normal human subjects in vivo. Am. Rev. Respir. Dis. 124:484–490.

Knowles, M. R., W. H. Buntin, P. A. Bromberg, J. T. Gatzy, and R. C. Boucher. 1982. Measurements of transepithelial electric potential differences in the trachea and bronchi of human subjects in vivo. Am. Rev. Respir. Dis. 126:108–112.

Knowles, M. R., J. T. Gatzy, and R. C. Boucher. 1983. Relative ion permeability of normal and cystic fibrosis nasal epithelium. J. Clin. Invest. 71:1410–1417.

Knowles, M. R., G. Murray, J. Shallal, F. Askin, V. Ranga, J. Gatzy, and R. Boucher. 1984. Bioelectric properties and ion flow across excised human bronchi. J. Appl. Physiol. 56:868–877.

Knowles, M. R., J. T. Gatzy, and R. C. Boucher. 1986. Measurement of canine tracheal epithelial resistance in vivo. Am. Rev. Respir. Dis. 133:A90. (Abstract.)

Koenig, J. Q., and W. E. Pierson. 1984. Nasal responses to air pollutants. Clin. Rev. Allergy 2:255–261.

Koenig, J. Q., D. S. Covert, S. G. Marshall, G. Van Belle, and W. E. Pierson. 1987. The effects of ozone and nitrogen dioxide on pulmonary function in healthy and in asthmatic adolescents. Am. Rev. Respir. Dis. 136:1152–1157.

Koponen, M., T. Gustafsson, K. Kalliomaki, P. L. Kalliomaki, M. Moilanen, and L. Pyy. 1980. Dusts in a steel making plant: Lung contamination among iron workers. Int. Arch. Occup. Environ. Health 47:35–45.

Koren, H. S., R. B. Devlin, D. E. Graham, R. Mann, M. P. McGee, D. H. Horstman, W. J. Kozumbo, S. Becker, D. E. House, W. F. McDonnell, and P. A. Bromberg. 1989. Ozone-induced inflammation in the lower airways of human subjects. Am. Rev. Respir. Dis. 139:407–415.

Koretzky, G. A., J. A. Elias, S. L. Kay, M. D. Rossman, P. C. Nowell, and R. P. Daniele. 1983. Spontaneous production of interleukin-1 by human alveolar macrophages. Clin. Immunol. Immunopathol. 29:443–450.

Krotoszynski, B., G. Gabriel, and H. O'Neill. 1977. Characterization of human expired air: A promising investigative and diagnostic technique. J. Chromatogr. Sci. 15:239–244.

Kulle, T. J., S. A. J. Goings, L. R. Sauder, D. J. Green, and M. L. Clements. 1987. Susceptibility to virus infection with NO$_2$ exposure. Am. Rev. Respir. Dis. 135(4 Pt. 2):A58. (Abstract.)

LaBelle, C. W., D. M. Bevilacqua, and H. Brieger. 1964. Synergistic effects of aerosols. IV. Therapeutic elimination of inhaled radioactive particles. J. Occup. Med. 6:391–395.

Lam, S., F. Tan, H. Chan, and M. Chan-Yeung. 1983. Relationship between types of asthmatic reaction, nonspecific bronchial reactivity, and specific IgE antibodies in patients with red cedar asthma. J. Allergy Clin. Immunol. 72:134–139.

Lambré, C. R., S. Le Maho, G. Di Bella, H. De Cremous, K. Atassi, and J. Bignon. 1986. Bronchoalveolar lavage fluid and serum complement activity in pulmonary sarcoidosis. Am. Rev. Respir. Dis. 134:238–247.

Lamontagne, L., J. Gauldie, A. Stadnyk, C. Richards, and E. Jenkins. 1985. In vivo initiation of unstimulated in vitro interleukin-1 release by alveolar macrophages. Am. Rev. Respir. Dis. 131:326–330.

Landahl, H. D. and R. G. Hermann. 1950. Retention of vapors and gases in the human nose and lung. AMA Arch. Ind. Health 1:36–45.

Lapp, N. L., J. L. Hankinson, H. Amandus, and E. D. Palmes. 1975. Variability in the size of airspaces in normal human lungs as estimated by aerosols. Thorax 30:293–299.

Larsen, G. L., B. C. Mitchell, T. B. Harper, and P. M. Henson. 1982. The pulmonary response of C5 sufficient and deficient mice to Pseudomonas aeruginosa. Am. Rev. Respir. Dis. 126:306–311.

Laurenzi, G. A., and J. J. Guarneri. 1966. Effects of bacteria and viruses on ciliated epithelium. A study of the mechanisms of pulmonary resistance to infection: The relationship of bacterial clearance to ciliary and alveolar macrophage function. Am. Rev. Respir. Dis. 93S:134–141.

Lauwreys, R. R. 1983. Industrial Chemical Exposure: Guidelines for Biological Monitoring. Davis, Calif.: Biomedical Publications. 150 pp.

Laviolette, M. 1985. Lymphocyte fluctuation in bronchoalveolar lavage fluid in normal volunteers. Thorax 40:651–656.

Laviolette, M., J. Chang, and D. S. Newcombe. 1981. Human alveolar macrophages: A lesion in arachidonic acid metabolism in cigarette smokers. Am. Rev. Respir. Dis. 124:397–401.

Leatherman, J. W., A. F. Michael, B. A. Schwartz, and J. R. Hoidal. 1984. Lung T cells in hypersensitivity pneumonitis. Ann. Intern. Med. 100:390–392.

Lee, C. T., A. M. Fein, M. Lippmann, H. Holtzman, P. Kimbel, and G. Weinbaum. 1981. Elastolytic activity in pulmonary lavage fluid from patients with adult respiratory-distress syndrome. N. Engl. J. Med. 304:192–196.

Lehnert, B. E., J. E. London, A. J. Vanderkogel, J. G. Valdez, and L. R. Gurley. 1986. HPLC analyses of lavage fluid constituents following environmental insults. Toxicologist 6:138. (Abstract.)

Leikauf, G., D. B. Yeates, K. A. Wales, D. Spektor, R. E. Albert, and M. Lippmann. 1981. Effects of sulfuric acid aerosol on respiratory mechanics and mucociliary particle clearance in healthy non-smoking adults. Am. Ind. Hyg. Assoc. J. 42:273–282.

Leith, D. E., and J. Mead. 1974. Principles of Body Plethysmography. [Bethesda, Md.:] Division of Lung Diseases, National Heart and Lung Institute.

Lem, V. M., M. F. Lipscomb, J. C. Weissler, G. Nunez, E. J. Ball, P. Stastny, and G. B. Toews. 1985. Bronchoalveolar cells from sarcoid patients demonstrate enhanced antigen presentation. J. Immunol. 135:1766–1771.

Le Mesurier, S. M., B. W. Steward, and A. W. J. Lykke. 1981. Injury to type-2 pneumocytes in rats exposed to cigarette smoke. Environ. Res. 24:207–217.

Leslie, C., K. McCormick-Shannon, J. Cook, and R. Mason. 1985. Macrophages stimulate DNA synthesis in rat alveolar Type II cells. Am. Rev. Respir. Dis. 132:1246.

Lewis, F. F., V. B. Elings, S. L. Hill, and J. M. Christensen. 1982. The measurement of extravascular lung water by thermal-green dye indicator dilution. Ann. N.Y. Acad. Sci. 384:394–410.

Lewis, R. A. 1985. A presumptive role for leukotrienes in obstructive airways diseases. Chest 88 (Suppl. 2):98S–102S.

Lewis, R. A., and K. F. Austen. 1984. The biologically active leukotrienes: Biosynthesis, metabolism, receptors, functions and pharmacology. J. Clin. Invest. 73:889–897.

Lima, M., and R. E. Rocklin. 1981. Histamine modulates in vitro IgG production by pokeweed mitogen-stimulated human mononuclear cells. Cell. Immunol. 64:324–336.

Lippmann, M. 1988. Rapporteur's Report. Session XI – U.S.-Dutch Ozone Symposium, May 11, 1988. [U.S.-Dutch Ozone Meeting, held May 9–13, 1988 in Nijmegen, The Netherlands.] 5 pp.

Lippmann, M., and R. B. Schlesinger. 1984. Interspecies comparisons of particle deposition and mucociliary clearance in tracheobronchial airways. J. Toxicol. Environ. Health 13:441–469.

Lippmann, M., D. B. Yeates, and R. E. Albert. 1980. Deposition, retention, and clearance of inhaled particles. Br. J. Ind. Med. 37:337–362.

Little, C. D., M. M. Nau, D. N. Carney, A. F. Gazdar, and J. D. Minna. 1983. Amplification and expression of the c-myc oncogene in human lung cancer cell lines. Nature 306:194–196.

Little, J. W., W. J. Hall, R. G. Douglas, Jr., G. S. Mudholkar, D. M. Speers, and K. Patel. 1978. Airway hyperreactivity and peripheral airway dysfunction in influenza A infection. Am. Rev. Respir. Dis. 118:295–303.

Liu, M. C., D. Proud, R. P. Schleimer, and M. Plaut. 1984. Human lung macrophages enhance and inhibit lymphocyte proliferation. J. Immunol. 132:2895–2903.

Loo, B. W., F. S. Gouldings, and D. S. Simon. 1986. A new compton densitometer for measuring pulmonary edema. IEEE Trans. Nucl. Sci. 33:531–536.

Lorin, M. I., P. F. Gaerlan, and I. D. Mandel. 1972. Quantitative composition of nasal secretions in normal subjects. J. Lab. Clin. Med. 80:275–281.

Loudon, R., and R. L. H. Murphy, Jr. 1984. Lung sounds. Am. Rev. Respir. Dis. 130:663–673.

Lourenco, R. V., E. D. Stanley, B. Gatmaiton, and G. G. Jackson. 1971a. Abnormal deposition and clearance of inhaled particles during upper respiratory viral infections. J. Clin. Invest. 50:62a. (Abstract.)

Lourenco, R. V., M. F. Klimek, and C. J. Borowski. 1971b. Deposition and clearance of 2-μ particles in the tracheobronchial tree of normal subjects-smokers and non-smokers. J. Clin. Invest. 50:1411–1419.

Low, R. B., G. S. Davis, and M. S. Giancola. 1978. Biochemical analysis of bronchoalveolar lavage fluids of healthy human volunteer smokers and nonsmokers. Am. Rev. Respir. Dis. 118:863–875.

Lucas, R. M., V. G. Iannacchione, and D. K. Melroy. 1982. Polychlorinated Biphenyls in Human Adipose Tissue and Mother's Milk. Washington, D.C.: U.S. Environmental Protection Agency.

Luchsinger, P. C., B. LaGarde, and J. E. Kilfeather. 1968. Particle clearance from the human tracheobronchial tree. Am. Rev. Respir. Dis. 97:1046–1050.

Luster, M. I., A. E. Munson, P. T. Thomas, M. P. Holsapple, J. D. Fenters, K. L. White, Jr., L. D. Lauer, D. R. Germolec, G. J. Rosenthal, and J. H. Deans. 1988. Development of a testing battery to assess chemical-induced immunotoxicity: National Toxicology Program's guidelines for immunotoxicity evaluation in mice. Fundam. Appl. Toxicol. 10:2-19.

Maccia, C. A., I. L. Bernstein, E. A. Emmett, and S. M. Brooks. 1976. In vitro demonstration of specific IgE phthalic anhydride hypersensitivity. Am. Rev. Respir. Dis. 113:701–704.

MacGlashan, D., Jr., R. P. Schleimer, J. S. Peters, E. S. Schulman, G. K. Adams, A. K. Sobotka, H. Newball, and L. M. Lichtenstein. 1983. Comparative studies of human basophils and mast cells. Fed. Prod. 42:2504–2509.

Macklem, P. T. 1974. Procedures For Standardized Measurements of Lung Mechanics. Measurement of Esophageal Pressure. Measurement of Lung Mechanics. [Bethesda, Md.:] Divison of Lung Diseases, National Heart and Lung Institute.

Macklem, P. T., and M. R. Becklake. 1963. The relationship between the mechanical and diffusing properties of the lung in health and disease. Am. Rev. Respir. Dis. 87:47–56.

Macklin, C. C. 1955. Pulmonary sumps, dust accumulations, alveolar fluid, and lymph vessels. Acta Anat. (Basal) 23:1–21.

Majima, Y., Y. Sakakura, T. Matsubara, S. Murai, and Y. Miyoshi. 1983. Mucociliary clearance in chronic sinusitis: Related human nasal clearance and in vitro bullfrog palate clearance. Biorheology 20:251–262.

Malm, L., J. Wihl, C. Lamm, and N. Lindqvist. 1981. Reduction of metacholine–induced nasal secretion by treatment with a new topical steroid in perennial non-allergic rhinitis. Allergy 36:209–214.

Malmberg H., and E. Holopainen. 1979. Nasal smear as a screening test for immediate-type nasal allergy. Allergy 34:331–337.

Man, S. F. P., T. K. Lee, R. T. N. Gibney, J. W. Logus, and A. A. Noujaim. 1980. Canine tracheal mucus transport of particulate pollutants: Comparison of radiolabeled corn pollen, ragweed pollen, asbestos, silica, and talc to Dowex anion exchange particles. Arch. Environ. Health 35:283–286.

Marcy, T. W., W. W. Merrill, J. A. Rankin, and H. Y. Reynolds. 1987. Limitations of using urea to quantify epithelial lining fluid recovered by bronchoalveolar lavage. Am. Rev. Respir. Dis. 135:1276–1280.

Marom, Z., J. Shelhamer, and M. Kaliner. 1984. Nasal mucus secretion. Ear Nose Throat J. 63:36–7, 41–2, 44.

Martin, B. M., M. A. Gimbrone, Jr., E. R. Unanue, and R. S. Coltran. 1981. Stimulation of nonlymphoid mesenchymal cell proliferation by a macrophage-derived growth factor. J. Immunol. 126:1510–1515.

Martin, R., and P. T. Macklem. 1973. Suggested Standardized Procedures for Closing Volume Determinations (Nitrogen Method). Bethesda, Md.: Division of Lung Diseases, National Heart and Lung Institute. 7 pp.

Martin, R. R., D. Lindsay, P. Despas, D. Bruce, M. Leroux, N. R. Anthonisen, and P. T. Macklem. 1975. The early detection of airway obstruction. Am. Rev. Respir. Dis. 111:119–125.

Martin, T. R., G. Raghu, R. J. Maunder, and S. C. Springmeyer. 1985. The effects of chronic bronchitis and chronic air-flow obstruction on lung cell populations recovered by bronchoalveolar lavage. Am. Rev. Respir. Dis. 132:254–260.

Martin, W. J., 2d, J. E. Gadek, G. W. Hunnighake, and R. G. Crystal. 1981. Oxidant injury of lung parenchymal cells. J. Clin. Invest. 68:1277–1288.

Mason, G. R., J. M. Uszler, R. M. Effros, and E. Reid. 1983. Rapidly reversible alterations of pulmonary epithelial permeability induced by smoking. Chest 83:6–11.

Mason, M. J., D. E. Bice, and B. A. Muggenburg. 1985. Local pulmonary immune responsiveness after multiple antigenic exposures in the cynomolgus monkey. Am. Rev. Respir. Dis. 132:657–660.

Mauderly, J. L. 1977. Bronchopulmonary lavage of small laboratory animals. Lab. Anim. Sci. 27:255–261.

Mauderly, J. L., B. A. Muggenburg, F. F. Hahn, and B. B. Boecker. 1980. The effects of inhaled 144Ce on cardiopulmonary function and histopathology of the dog. Radiat. Res. 84:307–324.

McAllen, S. J., S. P. Chiu, R. F. Phalen, and R. E. Rasmussen. 1981. Effect of in vivo ozone exposure on in vitro pulmonary alveolar macrophage mobility. J. Toxicol. Environ. Health 7:373–381.

McCarthy, D. S., R. Spencer, R. Greene, and J. Milic-Emili. 1972. Measurement of "closing volume" as a simple and sensitive test for early detection of small airway disease. Am. J. Med. 52:747–753.

McCawley, M. 1987. An aerosol bolus dispersion test compared to indices of lung disease. Ph.D. Dissertation, New York University, N.Y.

McCawley, M., and M. Lippmann. 1984. An aerosol dispersion test: Comparison of results from smokers and non-smokers, pp. 1007–1010. In B. Y. H. Liu, D. Y. H. Pui, and H. J. Fissan, Eds. Aerosols: Science, Technology and Industrial Applications of Airborne Particles. Proceedings of the First International Aerosol Conference held September 17–21, 1984, Minneapolis, Minnesota. New York: Elsevier.

McClellan, R. O. 1986. Health effects of diesel exhaust: A case study in risk assessment. Am. Ind. Hyg. Assoc. J. 47:1–13.

McCormick, J. R., and J. D. Catravas. 1986. Changes in the kinetics of pulmonary angiotensin-converting enzyme (ACE) and 5'-nucleotidase (NCT) following endothelial cell injury with phorbol myristate acetate (PMA). Am. Rev. Respir. Dis. 133:A19. (Abstract.)

McKay, R. T. 1986. Bronchoprovocation challenge testing in occupational airways disorders. Semin. Respir. Med. 7:297–306.

McLean, J. A., A. A. Ciarkowski, W. R. Solomon, and K. P. Mathews. 1976. An improved technique for nasal inhibition challenge tests. J. Allergy Clin. Immunol. 57:153–163.

McMahon, T. A., J. D. Brain, and S. LeMott. 1977. Species difference in aerosol deposition, pp. 23–33. In W. H. Walton, Ed. Inhaled Particles IV, Pt. 1. Oxford: Pergamon.

Mead, J., and E. Agostoni. 1964. Dynamics of breathing, pp. 411–427. In W. O. Fenn and H. Rahn, Eds. Handbook of Physiology, Section 3, Volume 1, Respiration. Bethesda, Md.: American Physiological Society.

Mecham, R. P., L. A. Whitehouse, D. S. Wrenn, W. C. Parks, G. L. Griffin, R. M. Senior, E. C. Crouch, K. R. Stenmark, and N. F. Voelkel. 1987. Smooth muscle-mediated connective tissue remodeling in pulmonary hypertension. Science 237(4813):423–426.

Medinsky, M. A., J. A. Dutcher, J. A. Bond, R. F. Henderson, J. L. Mauderly, M. B. Snipes, J. A. Mewhinney,

Y. S. Cheng, and L. S. Birnbaum. 1985. Uptake and excretion of [14C] methyl bromide as influenced by exposure concentration. Toxicol. Appl. Pharmacol. 78:215–225.

Medinsky, M. A., P. J. Sabourin, G. Lucier, L. S. Birnbaum, and R. F. Henderson. In press a. A physiological model for simulation of benzene metabolism by rats and mice. Toxicol. Appl. Pharmacol.

Medinsky, M. A., P. J. Sabourin, R. F. Henderson, G. W. Lucier, and L. S. Birnbaum. In press b. Differences in the pathways for metabolism of benzene in rats and mice simulated by a physiological model. Environ. Health Perspect. 82.

Merrill, W. W., J. G. Bartlett, and H. Y. Reynolds. 1982. Applied immunology of the lung, pp. 167–188. In D. H. Simmons, Ed. Current Pulmonology, Vol. 4. New York: John Wiley & Sons.

Metcalfe, D. D., M. Kaliner, and M. A. Donlon. 1981. The mast cell. CRC Crit. Rev. Immunol. 3:23–74.

Metzger, W. J., K. Nugent, H. B. Richerson, P. Moseley, R. Lakin, D. Zavala, and G. W. Hunninghake. 1985. Methods for bronchoalveolar lavage in asthmatic patients following bronchoprovocation and local antigen challenge. Chest 87(Suppl. 1):16S–19S.

Metzger, W. J., D. Zavala, H. B. Richerson, P. Moseley, P. Iwamota, M. Monick, K. Sjoerdsma, and G. W. Hunninghake. 1987. Local allergen challenge and bronchoalveolar lavage of allergic asthmatic lungs: Description of the model and local airway inflammation. Am. Rev. Respir. Dis. 135:433–440.

Meyrick, B., and K. L. Brigham. 1986. Repeated Escherichia coli endotoxin-induced pulmonary inflammation causes chronic pulmonary hypertension in sheep. Structural and functional changes. Lab. Invest. 55:164–176.

Meyrick, B., and L. Reid. 1978. The effect of continued hypoxia on rat pulmonary arterial circulation. An ultrastructural study. Lab. Invest. 38:188–200.

Meyrick, B., and L. Reid. 1979a. Ultrastructural features of the distended pulmonary arteries of the normal rat. Anat. Rec. 193:71–97.

Meyrick, B., and L. Reid. 1979b. Hypoxia and incorporation of 3H-thymidine by cells of the rat pulmonary arteries and alveolar wall. Am. J. Pathol. 96:51–70.

Meyrick, B., and L. Reid. 1980. Endothelial and subintimal changes in rat hilar pulmonary artery during recovery from hypoxia. A quantitative ultrastructural study. Lab. Invest. 42:603–615.

Meyrick, B., and L. Reid. 1983. Pulmonary hypertension. Anatomic and physiologic correlates. Clin. Chest Med. 4:199–217.

Meyrick, B., W. Gamble, and L. Reid. 1980. Development of Crotalaria pulmonary hypertension: Hemodynamic and structural study. Am. J. Physiol. 239:H692–H702.

Meyrick, B., E. A. Perkett, and K. L. Brigham. 1987. Inflammation and models of chronic pulmonary hypertension. Am. Rev. Respir. Dis. 136:765–767.

Mezey, R. J., M. A. Cohn, R. J. Fernandez, A. J. Januszkiewicz, and A. Wanner. 1978. Mucociliary transport in allergic patients with antigen-induced bronchospasm. Am. Rev. Respir. Dis. 118:677—684.

Miadonna, A., E. Leggieri, A. Tedeschi, and C. Zanussi. 1983. Clinical significance of specific IgE determination on nasal secretion. Clin. Allergy 13:155—164.

Miller, B. E., R. E. Chapin, L. B. Gilmore, K. E. Pinkerton, and G. E. R. Hook. 1986. Silica induced proliferation of alveolar Type II cells: Quantitation by alkaline phosphatase histochemistry. Toxicologist 6:133. (Abstract.)

Miller, B. E., R. E. Chapin, K. E. Pinkerton, L. B. Gilmore, R. R. Maronpot, and G. E. R. Hook. 1987. Quantitation of silica-induced type II cell hyperplasia by using alkaline phosphatase histochemistry in glycol methacrylate embedded lung. Exp. Lung Res. 12:135—148.

Miller, F. J., J. W. Illing, and D. E. Gardner. 1978. Effect of urban ozone levels on laboratory-induced respiratory infections. Toxicol. Lett. 2:163—169.

Miller, F. J., J. H. Overton, E. T. Myers, and J. A. Graham. 1982. Pulmonary dosimetry of nitrogen dioxide in animals and man, pp. 377—386. In T. Schneider and L. Grant, Eds. Air Pollution by Nitrogen Oxides: Proceedings of U.S.-Dutch International Symposium, Maastricht, The Netherlands. Studies in Environmental Science, 21. Amsterdam: Elsevier.

Miller, F. J., J. H. Overton, Jr., R. H. Jaskot, and D. B. Menzel. 1985. A model of the regional uptake of gaseous pollutants in the lung. I. The sensitivity of the uptake of ozone in the human lung to lower respiratory tract secretions and exercise. Toxicol. Appl. Pharmacol. 79:11—27.

Miller, F. J., J. H. Overton, Jr., E. D. Smolko, R. C. Graham, and D. B. Menzen. 1987. Hazard assessment using an integrated physiologically-based dosimetry modeling approach: Ozone, pp. 353—368. In Pharmacokinetics in Risk Assessment. Drinking Water and Health, Vol. 8. Washington, D.C.: National Academy Press.

Miller, R. E., J. L. Paradise, G. A. Friday, P. Fireman, and D. Voith. 1982. The nasal smear for eosinophils: Its value in children with seasonal allergic rhinitis. Am. J. Dis. Child. 136:1009—1011.

Milne, E. N. C., M. Pistolesi, M. Miniati, and C. Guintini. 1985. The radiologic distinction of cardiogenic and noncardiogenic edema. Am. J. Roentgen. 144:879—894.

Mintun, M. A., D. R. Dennis, M. J. Welch, C. J. Mathias, and D. P. Schuster. 1987. Measurements of pulmonary vascular permeability with PET and gallium-68 transferrin. J. Nucl. Med. 28:1704—1716.

Minty, B. D., and D. Royston. 1985. Cigarette smoke induced changes in rat pulmonary clearance of ^{99m}Tc-DTPA: A comparison of particulate and gas phases. Am. Rev. Respir. Dis. 132:1170—1173.

Mishra, B. B., L. W. Poulter, G. Janossy, S. Sherlock, and D. G. James. 1986. The Kveim-Siltzbach granuloma: A model for sarcoid granuloma formation. Ann. N.Y. Acad. Sci. 465:164—175.

Miskulin, M., R. Dalgleish, B. Kluve-Beckerman, S. I. Rennard, P. Tolstoshev, M. Brantly, and R. G. Crystal. 1986. Human type III collagen gene expression is coordinately modulated with the type I collagen genes during fibroblast growth. Biochemistry 25:1408—1413.

Mizel, S. B. 1982. Interleukin 1 and T cell activation. Immunol. Rev. 63:51—72.

Mohtashamipur, E., K. Norpoth, and F. Lieder. 1985. Isolation of frameshift mutagens from smokers' urine: Experience with three concentration methods. Carcinogenesis 6:783—788.

Monaco, A. P., R. L. Neve, C. Colletti-Feener, C. J. Bertelson, D. M. Kurnit, and L. M. Kunkel. 1986. Isolation of candidate cDNAs for portions of the Duchenne muscular dystrophy gene. Nature 323:646—650.

Monaco, A. P., C. J. Bertelson, C. Colletti-Feener, and L. M. Kunkel. 1987. Localization and cloning of Xp21 deletion breakpoints involved in muscular dystrophy. Hum. Genet. 75:221—227.

Monick, M., J. Glazier, and G. W. Hunninghake. 1987. Human alveolar macrophages suppress interleukin-1 (Il-1) activity via the secretion of prostaglandin E_{22}. Am. Rev. Respir. Dis. 135:72—77.

Moore, V. L., G. M. Pedersen, W. C. Hauser, and J. N. Fink. 1980. A study of lung lavage materials in patients with hypersensitivity pneumonitis: In vitro response to mitogen and antigen in pigeon breeders' disease. J. Allergy Clin. Immunol. 65:365—370.

Moores, S. R., S. E. Sykes, A. Morgan, N. Evans, J. C. Evans, and A. Holmes. 1980. The short-term cellular and biochemical response of the lung to toxic dusts: An in vivo cytotoxicity test, pp. 297—303. In R. C. Brown, I. P. Gormley, M. Chamberlain, and R. Davies, Eds. The In Vitro Effect of Mineral Dusts. New York: Academic Press.

Morgan, A., A. Black, J. C. Evans, E. H. Hadfield, R. G. Macbeth, and M. Walsh. 1973. Impairment of nasal mucociliary clearance in wood workers in the furniture industry, pp. 335—338. In Proceedings of First International Congress on Aerosols in Medicine, 19—21 September 1973, Baden/Wien, Austria. Baden/Wien: Internationale Gesellschaft für Aerosole in der Medizin.

Morgan, A., S. R. Moores, A. Holmes, J. C. Evans, N. H. Evans, and A. Black. 1980. The effect of quartz, administered by intratracheal instillation, on the rat lung. I. The cellular response. Environ. Res. 22:1—12.

Morley, A. A., K. J. Trainor, R. Seshadri, and R. G. Ryall. 1983. Measurement of in vivo mutations in human lymphocytes. Nature 302:155—156.

Mornex, J. F., Y. Martinet, K. Yamauchi, P. B. Bitterman, G. R. Grotendorst, A. Chytil-Weir, G. R. Martin, and R. G. Crystal. 1986. Spontaneous expression of the c-sis gene and release of a platelet-derived growth factor like molecule by human alveolar macrophages. J. Clin. Invest. 78:61—66.

Morris, A. H., D. D. Blatter, T. A. Case, A. G. Cutillo, D. C. Ailion, C. H. Durney, and S. A. Johnson. 1985.

A new nuclear magnetic resonance property of lung. J. Appl. Physiol. 58:759—762.

Morris, J. B., and D. G. Cavanagh. 1986. Deposition of ethanol and acetone vapors in the upper respiratory tract of the rat. Fund. Appl. Toxicol. 6:78—88.

Morris, J. B., R. J. Clay, and D. G. Cavanagh. 1986. Species differences in upper respiratory tract deposition of acetone and ethanol vapors. Fundam. Appl. Toxicol. 7:671—680.

Mossberg, B., and P. Camner. 1980. Impaired mucociliary transport as a pathogenetic factor in obstrutive pulmonary diseases. Chest 77(Suppl. 2):265—266.

Muggenburg, B. A., J. L. Mauderly, J. A. Pickrell, T. L. Chifelle, R. K. Jones, U. C. Luft, R. O. McClellan, and R. C. Pfleger. 1972. Pathophysiologic sequelae of bronchopulmonary lavage in the dog. Am. Rev. Respir. Dis. 106:219—232.

Muggenburg, B. A., F. F. Hahn, J. A. Bowen, and D. E. Bice. 1982. Flexible fiberoptic bronchoscopy of chimpanzees. Lab. Anim. Sci. 32:534—537.

Muir, D. C. F. 1970. The effect of airways obstruction on the single breath aerosol curve, pp. 319—325. In A. Bouhuys, Ed. Airway Dynamics: Physiology and Pharmacology. Springfield, Ill.: Charles C Thomas.

Muller-Quernheim, J., C. Saltini, P. Sondermeyer, and R. G. Crystal. 1986. Compartmentalized activation of the interleukin 2 gene by lung T lymphocytes in active pulmonary sarcoidosis. J. Immunol. 137:3475—3483.

Mundie, T. G., C. Whitener, and S. K. Ainsworth. 1985. Byssinosis: Release of prostaglandins, thromboxanes, and 5-hydroxytryptamine in bronchopulmonary lavage fluid after inhalation of cotton dust extracts. Am. J. Pathol. 118:128—133.

Murray, J. J., A. B. Tonnel, A. R. Brash, L. J. Roberts, 2d, P. Gosset, R. Workman, A. Capron, and J. A. Oates. 1986. Release of prostaglandin D_2 into human airways during acute antigen challenge. N. Engl. J. Med. 315:800—804.

Mygind, N., B. Weeke, and S. Ullman. 1975. Quantitative determination of immunoglobulins in nasal secretion. Int. Arch. Allergy Appl. Immunol. 49:99—107.

Naclerio, R. M., H. L. Meier, A. Kagey-Sobotka, N. F. Adkinson, Jr., D. A. Meyers, P. S. Norman, and L. M. Lichtenstein. 1983. Mediator release after nasal airway challenge with allergen. Am. Rev. Respir. Dis. 128:597—602.

Naclerio, R. M., D. Proud, A. G. Togias, N. F. Adkinson, Jr., D. A. Meyers, A. Kagey-Sobotka, M. Plaut, P. S. Norman, and L. M. Lichtenstein. 1985. Inflammatory mediators in late antigen-induced rhinitis. N. Engl. J. Med. 313:65—70.

Nadel, J. A., and P. J. Barnes. 1984. Autonomic regulation of the airways. Annu. Rev. Med. 34:451—467.

Nebert, D. W., and F. J. Gonzalez. 1987. P450 genes: Structure, evolution, and regulation. Annu. Rev. Biochem. 56:945—993.

Neiman, P. E., W. Reeves, G. Ray, N. Flournoy, K. G. Lerner, G. E. Sale, and E. D. Thomas. 1977. A prospective analysis of interstitial pneumonia and opportunistic viral infection among recipients of allogeneic bone marrow grafts. J. Infect. Dis. 136:754—767.

Nemery, B., N. E. Moavero, L. Brasseur, and D. C. Stanescu. 1981. Significance of small airway tests in middle-aged smokers. Am. Rev. Respir. Dis. 124:232—238.

Newmann, H.-G. 1984. Analysis of hemoglobin as a dose monitor for alkylating and arylating agents. Arch. Toxicol. 56:1—6.

Newman Taylor, A. J., J. R. Myers, J. L. Longbottom, D. Spackman, and A. J. M. Slovak. 1981. Immunological differences between asthma and other allergic reactions in laboratory animal workers. Thorax 36:229. (Abstract.)

NHLBI (National Heart, Lung and Blood Institute). 1985. Summary and recommendations of a workshop on the investigative use of fiberoptic bronchoscopy and bronchoalveolar lavage in asthmatics. Am. Rev. Respir. Dis. 132:180—182.

NHLI (National Heart and Lung Institute). 1973. Workshop on screening programs for early diagnosis of airway obstruction. October. Bethesda, Md.: National Heart and Lund Institute, Division of Lung Diseases.

Niederman, M. S., T. D. Rafferty, C. T. Sasaki, W. W. Merrill, R. A. Matthay, and H. Y. Reynolds. 1983. Comparison of bacterial adherence to cilated and squamous epithelial cells obtained from the human respiratory tract. Am. Rev. Respir. Dis. 127:85-90.

Nikiforov, A. I., and R. B. Schlesinger. 1985. Morphometric variability of the human upper bronchial tree. Respir. Physiol. 59:289-299.

Nolop, K. B., D. L. Maxwell, D. Royston, and J. M. Hughes. 1986. Effect of raised thoracic pressure and volume on ^{99m}Tc-DTPA clearance in humans. J. Appl. Physiol. 60:1493-1497.

Norman, P. S. 1980. Skin testing, pp. 789—793. In N. R. Rose and H. Friedman, Eds. Manual of Clinical Immunology, 2d ed. Washington, D.C.: American Society for Microbiology.

Normand, I. C. S., R. E. Oliver, E. O. R. Reynolds, and L. B. Strang. 1971. Permeability of lung capillaries and alveoli to non-electrolytes in foetal lamb. J. Physiol. 219:303—330.

NRC (National Research Council). 1987. Pharmacokinetics in Risk Assessment. Drinking Water and Health, Vol. 8. Washington, D.C.: National Academy Press. 488 pp.

NRC (National Research Council). 1989. Drinking Water and Health, Vol. 9. Washington, D.C.: National Academy Press.

NTP (National Toxicology Program). 1986. Toxicology and Carcinogenesis Studies of Benzene in F344/N Rats and B6C3F$_1$ Mice (Gavage Studies). NTP TR 289. Research Triangle Park, N.C.: U.S. Department of Health and Human Services. 277 pp.

Nugent, K. M., J. Glazier, J., M. M. Monick, and G. W.

Hunninghake. 1985. Stimulated human alveolar macrophages secrete interferon. Am. Rev. Respir. Dis. 131:714—718.

O'Byrne, P. M. 1986. Airway inflammation and airway hyperresponsiveness. Chest 90:575—577.

O'Byrne, P. M., M. Dolovich, R. Dirks, R. S. Roberts, and M. T. Newhouse. 1984. Lung epithelial permeability: Relation to nonspecific airway responsiveness. J. Appl. Physiol. 57:77—84.

Oberdörster, G., and A. P. Freedman. 1981. Lung clearance measurements in rats using magnetopneumography, pp. 19—24. In W. Stober, and D. Hochrainer, Eds. Aerosols in Science, Medicine and Technology, Vol. 9. Schmallenberg, F.R.G.: Gesellschaft fur Aerosolforschung.

Oberdörster, G., M. J. Utell, P. E. Morrow, R. W. Hyde, and D. A. Weber. 1986. Bronchial and alveolar absorption of inhaled ^{99m}Tc-DTPA. Am. Rev. Respir. Dis. 134:944—950.

Ogilvie, C. M., R. E. Forster, W. S. Blakemore, and J. W. Morton. 1957. A standardized breath holding technique for the clinical measurement of the diffusing capacity of the lung for carbon monoxide. J. Clin. Invest. 36:1—17.

Ogra, P. L., J. C. Cumella, and R. C. Welliver. 1984. Immune response to viruses, pp. 242—263. In J. Bienenstock, Ed. Immunology of the Lung and Upper Respiratory Tract. New York: McGraw-Hill.

Ohyama, S., T. Inamasu, M. Ishizawa, N. Ishinishi, and K. Matsuura. 1987. Mutagenicity of human urine after the consumption of fried salted salmon. Food Chem. Toxicol. 25:147—153.

Osterman-Golkar, S., L. Ehrenberg, D. Segerback, and I. Hallstrom. 1976. Evaluation of genetic risks of alkylating agents. II. Haemoglobin as a dose monitor. Mutat. Res. 34:1—10.

Osterman-Golkar, S., P. B. Farmer, D. Segerback, E. Bailey, C. J. Calleman, K. Svensson, and L. Ehrenberg. 1983. Dosimetry of ethylene oxide in the rat by quantitation of alkylated histidine in hemoglobin. Teratogenesis Carcinog. Mutagen. 3:395—405.

Otis, A. B., C. B. McKerrow, R. A. Bartlett, J. Mead, M. B. McIlroy, N. J. Selverstone, and E. P. Radford. 1956. Mechanical factors in distribution of pulmonary ventilation. J. Appl. Physiol. 8:427—443.

Overland, E. S., R. N. Gupta, G. J. Huchon, and J. F. Murray. 1981. Measurement of pulmonary tissue volume and blood flow in persons with normal and edematous lungs. J. Appl. Physiol. 51:1375—1383.

Pacht, E. R., H. Kaseki, J. R. Mohammed, D. G. Cornwell, and W. B. Davis. 1986. Deficiency of vitamin E in the alveolar fluid of cigarette smokers. Influence on alveolar macrophage cytotoxicity. J. Clin. Invest. 77:789—796.

Palmes, E. D., B. Altshuler, and N. Nelson. 1967. Deposition of aerosols in the human respiratory tract during breath holding, pp. 339—347. In C. N. Davies, Ed. Inhaled Particles and Vapours II. Oxford: Pergamon.

Palmes, E. D., R. M. Goldring, C. S. Wang, and B. Altshuler. 1971. Effect of chronic obstructive pulmonary disease on rate of deposition of aerosols in the lung during breath holding, pp. 123—130. In W. H. Walton, Ed. Inhaled Particles III. Surry: Unwin.

Palmes, E. D., C. S. Wang, R. M. Goldring, and B. Altshuler. 1973. Effect of depth of inhalation on aerosol persistence during breath holding. J. Appl. Physiol. 34:356—360.

Paradis, I. L., J. H. Dauber, and B. S. Rabin. 1986. Lymphocyte phenotypes in bronchoalveolar lavage and lung tissue in sarcoidosis and idiopathic pulmonary fibrosis. Am. Rev. Respir. Dis. 133:855—860.

Patterson, D. G., R. E. Hoffman, L. L. Needham, D. W. Roberts, J. R. Bagby, J. L. Pirkle, H. Falk, E. J. Sampson, and V. N. Houk. 1986. 2,3,7,8-Tetrachlorodibenzo-p-dioxin levels in adipose tissue of exposed and control persons in Missouri: An interim report. J. Am. Med. Assoc. 256:2683—2686.

Patton, S. E., L. B. Gilmore, A. M. Jetten, P. Nettesheim, and G. E. Hook. 1986. Biosynthesis and release of proteins by isolated pulmonary Clara cells. Exp. Lung Res. 11:277—294.

Penn, I. 1978. Development of cancer in transplantation patients. Adv. Surg. 12:155—191.

Pepys, J., and B. J. Hutchcroft. 1975. Bronchial provocation tests in etiologic diagnosis and analysis of asthma. Am. Rev. Respir. Dis. 112:829—859.

Pereira, M. A., and L. W. Chang. 1981. Binding of chemical carcinogens and mutagens to rat hemoglobin. Chem.-Biol. Interact. 33:301—305.

Pereira, M. A., F. J. Burns, and R. E. Albert. 1979. Dose response for benzo(a)pyrene adducts in mouse epidermal DNA. Cancer Res. 39:2556—2559.

Perera, F. P., and I. B. Weinstein. 1982. Molecular epidemiology and carcinogen-DNA adduct detection: New approaches to studies of human cancer causation. J. Chronic Dis. 35:581—600.

Perez, H. D. 1984. Biologically active complement (C5)-derived peptides and their relevance to disease. CRC Crit. Rev. Oncol. Hematol. 1:199—225.

Perkett, E. A., K. L. Brigham, and B. Meyrick. 1988. Granulocyte depletion (PMN) attenuates chronic pulmonary hypertension following pulmonary air embolization. Fed. Am. Soc. Exp. Bio. J. 2:A1579. (Abstract.)

Peslin, R., and J. J. Fredberg. 1986. Oscillation mechanics of the respiratory system, pp. 145—177. In P. T. Macklem and J. Mead, Eds. Handbook of Physiology, Section 3: The Respiratory System. Volume III. Mechanics of Breathing, Part 1. Bethesda, Md.: American Physiological Society.

Peslin, R., C. Gallina, and C. Duvivier. 1986. Respiratory transfer impedances with pressure input at the mouth and chest. J. Appl. Physiol. 61:81—86.

Peterson, B. T., and L. D. Gray. 1987. Pulmonary lymphatic clearance of ^{99m}Tc-DTPA from the air spaces during lung inflation and lung injury. J. Appl. Physiol. 63:1136—1141.

Peterson, B. T., H. L. James, and J. W. McLarty. 1986. Effects of lung volume on the compartmental analysis of alveolar solute clearance in sheep. Physiologist 29:172. (Abstract.)

Peterson, B. T., H. L. James, and J. W. McLarty. 1988. Effects of lung volume on clearance of solutes from the air spaces of lungs. J. Appl. Physiol. 64:1068–1075.

Peterson, B. T., J. D. Dickerson, H. L. James, E. J. Miller, J. W. McLarty, and D. B. Holiday. 1989. Comparison of three tracers for detecting lung epithelial injury in anesthetized sheep. J. Appl. Physiol. 66:2374–2383.

Peto, R., F. E. Speizer, A. L. Cochrane, F. Moore, C. M. Fletcher, C. M. Tinker, I. T. T. Higgins, R. G. Gray, S. M. Richards, J. Gilliland, and B. Norman-Smith. 1983. The relevance in adults of air-flow obstruction, but not of mucus hypersecretion, to mortality from chronic lung disease. Am. Rev. Respir. Dis. 128:491–500.

Petty, T. L., G. W. Silvers, R. E. Stanford, M. D. Baird, and R. S. Mitchell. 1980. Small airway pathology is related to increased closing capacity and abnormal slope of phase III in excised human lungs. Am. Rev. Respir. Dis. 121:449–456.

Phalen, R. F., R. C. Mannix, and R. T. Drew. 1984. Inhalation exposure methodology. Environ. Health Perspect. 56:23–34.

Pickrell, J. A. 1981. Collagen degradation, pp. 61–70. In Lung Connective Tissue: Location, Metabolism and Response to Injury. Boca Raton, Fla.: CRC Press.

Pickrell, J. A., R. E. Gregory, D. J. Cole, F. F. Hahn, and R. F. Henderson. 1987. Effect of acute ozone exposure on the proteinase-antiproteinase balance in the rat lung. Exp. Mol. Pathol. 46:168–179.

Pinkston, P., P. B. Bitterman, and R. G. Crystal. 1983. Spontaneous release of interleukin-2 by lung T lymphocytes in active pulmonary sarcoidosis. N. Engl. J. Med. 308:793–800.

Pinkston, P., P. Bitterman, A. Hance, S. Douches, S. Goodman, and R. Crystal. 1984. In vivo demonstration of the importance of interleukin-2 in the proliferation of human T-lymphocytes. Clin. Res. 32:355A. (Abstract.)

Pirastu, M., Y. W. Kan, A. Cao, B. J. Conner, R. L. Teplitz, and R. B. Wallace. 1983. Prenatal diagnosis of beta-thalassemia: Detection of a single nucleotide mutation in DNA. N. Engl. J. Med. 309:284–287.

Pistolesi, M., and C. Guintini. 1978. Assessment of extravascular lung water. Radiol. Clin. North Am. 16:551–574.

Poirier, M. C., and F. A. Beland. 1987a. Determination of carcinogen-induced macromolecular adducts in animals and humans. Prog. Exp. Tumor Res. 31:1–10.

Poirier, M. C., and F. A. Beland, Eds. 1987b. Carcinogenesis and Adducts in Animals and Humans. Proceedings of the Workshop on Carcinogenesis and Adducts in Humans and Animals, Cambridge, Mass., Feb. 25, 1986. Progress in Experimental Tumor Research, Vol. 31. Basel: Karger. 117 pp.

Poirier, M. C., E. Reed, R. F. Ozols, T. Fasy, and S. H. Yuspa. 1987. DNA adducts of cisplatin in nucleated peripheral blood cells and tissues of cancer patients. Prog. Exp. Tumor Res. 31:104–113.

Powell, K. R., R. Shorr, J. D. Cherry, and J. O. Hendley. 1977. Improved method for collection of nasal mucus. J. Infect. Dis. 136:109–111.

Prochownik, E. V., S. Antonarakis, K. A. Bauer, R. D. Rosenberg, E. R. Fearon, and S. H. Orkin. 1983. Molecular heterogeneity of inherited antithrombin III deficiency. N. Engl. J. Med. 308:1549–1552.

Proctor, D. F. 1979. Tests of airway defense mechanisms, pp. 227–241. In P. E. Macklem and S. Permutt, Eds. The Lung in the Transition Between Health and Disease. Lung Biology in Health and Disease, Vol. 12. New York: Marcel Dekker.

Proctor, D. F. 1982. The upper airway, pp. 23–44. In D. F. Proctor and I. Andersen, Eds. The Nose: Upper Airway Physiology and the Atmospheric Environment. New York: Elsevier.

Proctor, D. F., I. B. Andersen, and G. Lundqvist. 1977. Nasal mucociliary function in humans, pp. 427–452. In J. D. Brain, D. F. Proctor and L. M. Reid, Eds. Respiratory Defense Mechanisms, Part 1. Lung Biology in Health and Disease, Vol. 5, Pt. 1. New York: Marcel Dekker.

Puchelle, E., F. Aug, J. M. Zahm, and A. Bertrand. 1982. Comparison of nasal and bronchial mucociliary clearance in young non-smokers. Clin. Sci. 62:13–16.

Quinlan, M. F., S. D. Salman, D. L. Swift, H. N. Wagner, Jr., and D. F. Proctor. 1969. Measurement of mucociliary function in man. Am. Rev. Respir. Dis. 99:13–23.

Rabinovitch, M., W. J. Gamble, O. S. Miettinen, and L. Reid. 1979. Age and sex influence on pulmonary hypertension of chronic hypoxia and on recovery. Am. J. Physiol. 240:H62–H72.

Rabinovitch, M., J. F. Keane, K. E. Fellows, A. R. Castaneda, and L. Reid. 1981. Quantitative analysis of the pulmonary wedge angiogram in congenital heart defects. Correlation with hemodynamic data and morphometric findings in lung biopsy tissue. Circulation 63:152–164.

Rabinovitch, M., J. F. Kean, W. I. Norwood, A. R. Castaneda, and L. Reid. 1984. Vascular structure in lung tissue obtained at biopsy correlated with pulmonary hemodynamic findings after repair of congenital heart defects. Circulation 69:655–657.

Ramanna, L., D. P. Tashkin, G. V. Taplin, D. Elam, R. Detels, A. Coulson, and S. N. Rokaw. 1975. Radioaerosol lung imaging in chronic obstructive pulmonary disease: Comparison with pulmonary function tests and roentgenography. Chest 68:634–640.

Ramsey, J. C., and M. E. Andersen. 1984. A physiologically based description of the inhalation pharmacokinetics of styrene in rats and humans. Toxicol. Appl. Pharmacol. 73:159–175.

Ramsey, J. C., J. D. Young, R. Karbowski, M. B. Chenoweth, L. P. McCarty, and W. H. Braun. 1980. Pharmacokinetics of inhaled styrene in human volunteers.

Toxicol. Appl. Pharmacol. 53:54—63.

Randerath, K., E. Randerath, H. P. Agrawal, R. C. Gupta, M. E. Schurdak, and M. V. Reddy. 1985. Postlabeling methods of carcinogen-DNA adduct analysis. Environ. Health Perspect. 62:57—65.

Ranga, V., J. Kleinerman, M. P. C. Ip, and A. M. Collins. 1980. The effect of nitrogen dioxide on tracheal uptake and transport of horseradish peroxidase in the guinea pig. Am. Rev. Respir. Dis. 122:483—490.

Rankin, J. A., P. E. Synder, E. N. Schachter, and R. A. Matthay. 1984. Bronchoalveolar lavage: Its safety in subjects with mild asthma. Chest 85:723—728.

Rao, A., S. B. Mizel, and H. Cantor. 1983. Disparate functional properties of two interleukin 1-responsive Ly-1+2-T cell clones: Distinction of T cell growth factor and T cell-replacing factor activities. J. Immunol. 130:1743—1748.

Razin, A., and A. D. Riggs. 1980. DNA methylation and gene function. Science 210:604—610.

Rearick, J., and A. M. Jetten. 1986. Accumulation of cholesterol 3-sulfate during in vitro squamous differentiation of rabbit tracheal epithelial cells and its regulation by retinoids. J. Biol. Chem. 261:13898—13904.

Rearick, J. I., M. Deas, and A. M. Jetten. 1987. Synthesis of mucous glycoproteins by rabbit tracheal cells in vitro: Modulation by substratum, retinoids and cyclic AMP. Biochem. J. 242:19—25.

Reasor, M. J., D. Nadeau, and G. E. R. Hook. 1978. Extracellular alkaline phosphatase in the airways of the rabbit lung. Lung 155:321—325.

Reddy, E. P., R. K. Reynolds, E. Santos, and M. Barbacid. 1982. A point mutation is responsible for the acquisition of transforming properties by the T24 bladder carcinoma oncogene. Nature 300:149—152.

Regal, J. F. 1988. Role of arachidonate metabolites in C5a-induced bronchoconstriction. J. Pharmacol. Exp. Ther. 246:542—547.

Reid, K. B., and E. Solomon. 1977. Biosynthesis of the first component of complement by human fibroblasts. Biochem. J. 167:647—660.

Reid, L. M. 1979. The pulmonary circulation: Remodeling in growth and disease. Am. Rev. Respir. Dis. 119:531—546.

Reitz, R. H., T. R. Fox, and P. G. Watanabe. 1986. The role of pharmacokinetics in risk assessment. Basic Life Sci. 38:499—507.

Remington, J. S., K. L. Vosti, A. Lietze, et al. 1964. Serum proteins and antibody activity in human nasal secretions. J. Clin. Invest. 43:1613—1624.

Rennard, S. I., and R. G. Crystal. 1982. Fibronectin in human bronchopulmonary lavage fluid. Elevation in patients with interstitial lung disease. J. Clin. Invest. 69:113—22.

Rennard, S. I., G. Basset, D. Lecossier, K. M. O'Donnell, P. Pinkston, P. G. Martin, and R. G. Crystal. 1986. Estimation of absolute volume of epithelial lining fluid recovered by lavage using urea as a marker of dilution. J. Appl. Physiol. 60:532—538.

Reynolds, H. Y. 1979. Lung host defenses: A status report. Chest 75(Suppl. 2):239—242.

Reynolds, H. Y. 1987. Bronchoalveolar lavage. Am. Rev. Respir. Dis. 135:250—263.

Reynolds, H. Y., and W. W. Merrill. 1981. Pulmonary immunology: Humoral and cellular immune responsiveness of the respiratory tract, pp. 381—422. In D. H. Simmons, Ed. Current Pulmonology, Vol. 3. New York: John Wiley & Sons.

Reynolds, H. Y., and H. H. Newball. 1974. Analysis of proteins and respiratory cells obtained from human lungs by bronchial lavage. J. Lab. Clin. Med. 84:559—573.

Reynolds, H. Y., J. A. Kazmierowski, and H. H. Newball. 1975. Specificity of opsonic antibodies to enhance phagocytosis of Pseudomonas aeruginosa by human alveolar macrophages. J. Clin. Invest. 56:376—385.

Reynolds, H. Y., J. D. Fulmer, J. A. Kazmierowski, W. C. Roberts, M. M. Frank, and R. G. Crystal. 1977. Analysis of cellular and protein content of broncho-alveolar lavage fluid from patients with idiopathic pulmonary fibrosis and chronic hypersensitivity pneumonitis. J. Clin. Invest. 59:165—175.

Rhodes, C. G., P. Wollmer, F. Fazio, and T. Jones. 1981. Quantitative measurement of regional extravascular lung density using positron emission and transmission tomography. J. Comput. Assist. Tomogr. 5:783—791.

Richter, A. M., R. T. Abboud, and S. S. Johal. 1986. Methylene blue decreases the functional activity of alpha-1-protease inhibitor in bronchoalveolar lavage. Am. Rev. Respir. Dis. 134:326—327.

Rinderknecht, J., L. Shapiro, M. Krauthammer, G. Taplin, K. Wasserman, J. M. Uszler, and R. M. Effros. 1980. Accelerated clearance of small solutes from the lungs in interstitial lung disease. Am. Rev. Respir. Dis. 121:105—117.

Rizk, N. W., J. M. Luce, J. M. Hoeffel, D. C. Price, and J. F. Murray. 1984. Site of deposition and factors affecting clearance of aerosolized solute from canine lungs. J. Appl. Physiol. 56:723—729.

Robbins, R. A., J. E. Gadek, and R. G. Crystal. 1981. Potential role of the complement system in propagating the alveolitis of idiopathic pulmonary fibrosis. Am. Rev. Respir. Dis. 123(4 Pt. 2):50. (Abstract.)

Roberts, J. A., I. W. Rodger, and N. C. Thomson. 1985. Airway responsiveness to histamine in men: Effect of atropine on in vivo and in vitro comparison. Thorax 40:261—267.

Robertson, J., J. R. Caldwell, J. R. Castle, and R. H. Waldman. 1976. Evidence for the presence of components of the alternative (properdin) pathway of complement activation in respiratory secretions. J. Immunol. 117:900—903.

Robins, R. A., J. E. Gadek, S. I. Rennard, Y. F. Chen, and R. G. Crystal. 1982. Compartmentalization of the complement system: The normal lung is C5 "deficient." Am. Rev. Respir. Dis. 125:53. (Abstract.)

Robinson, B. W., T. L. McLemore, and R. G. Crystal. 1985. Gamma interferon is spontaneously released by alveolar macrophages and lung T lymphocytes in patients with pulmonary sarcoidosis. J. Clin. Invest. 75:1488–1495.

Robinson, S. E., and A. P. Freedman. 1979. Direct magnetopneumographic measurement of particle concentrations and clearance in the lung, pp. 183–188. In Institute of Electrical and Electronics Engineers, Engineering in Medicine and Biology Society. Frontiers of Engineering in Health Care. New York: Institute of Electrical and Electronics Engineers.

Rodarte, J. R., and K. Rehder. 1986. Dynamics of respiration, pp. 131–144. In P. T. Macklem and J. Mead, Eds. Handbook of Physiology, Section 3: The Respiratory System. Volume III. Mechanics of Breathing, Part 1. Bethesda, Md.: American Physiological Society.

Roggli, V. L., and A. R. Brody. 1984. Changes in numbers and dimensions of chrysotile asbestos fibers in lungs of rats following short-term exposure. Exp. Lung Res. 7:133–147.

Roggli, V. L., C. A. Piantadosi, and D. Y. Bell. 1986. Asbestos bodies in bronchoalveolar lavage fluid: A study of 20 asbestos-exposed individuals and comparison to patients with other chronic interstitial lung diseases. Acta Cytol. 30:470–476.

Roggli, V. L., M. George, and A. R. Brody. 1987. Clearance and dimensional changes of crocidolite asbestos fibers isolated from lungs of rats following short-term exposure. Environ. Res. 42:94–105.

Rom, W. N., J. E. Lockey, K. M. Bang, C. Dewitt, and R. E. Johns, Jr. 1983. Reversible beryllium sensitization in a prospective study of beryllium workers. Arch. Environ. Health 38:302–307.

Roop, D. R., J. L. Nordstrom, S. Y. Tsai, M. J. Tsai, and B. W. O'Malley. 1978. Transcription of structural and intervening sequences in the ovalbumin gene and identification of potential ovalbumin mRNA precursors. Cell 15:671–685.

Rossen, R. D., R. H. Alford, W. T. Butler, and W. E. Vannier. 1966. The separation and characterization of proteins intrinsic to nasal secretion. J. Immunol. 97:369–378.

Rossen, R. D., J. A. Kasel, and R. B. Couch. 1971. The secretory immune system: Its relation to respiratory viral infection. Prog. Med. Virol. 13:194–238.

Rossi, G. A., G. B. Di Negro, E. Balzano, E. Cerri, O. Sacco, B. Balbi, A. Venturini, R. Ramoino, and C. Ravazzoni. 1985. Suppression of the alveolitis of pulmonary sarcoidosis by oral corticosteroids. Lung 163:83–93.

Rossman, C. M., R. M. K. W. Lee, J. B. Forrest, and M. T. Newhouse. 1984. Nasal ciliary ultrastructure and function in patients with primary ciliary dyskinesia compared with that in normal subjects and in subjects with various respiratory diseases. Am. Rev. Respir. Dis. 129:161–167.

Rossman, M. D., J. H. Dauber, and R. P. Daniele. 1978. Identification of activated T cells in sarcoidosis. Am. Rev. Respir. Dis. 117:713–720.

Roth, R. A. 1985. Biochemistry, physiology and drug metabolism: Implications regarding the role of the lungs in drug disposition. Clin. Physiol. Biochem. 3:66–79.

Sabourin, P. J., B. T. Chen, G. Lucier, L. S. Birnbaum, E. Fisher, and R. F. Henderson. 1987a. Effect of dose on the absorption and excretion of [^{14}C]benzene administered orally or by inhalation in rats and mice. Toxicol. Appl. Pharmacol. 87:325–336.

Sabourin, P. J., W. Bechtold, L. Birnbaum, G. Lucier, and R. Henderson. 1987b. Water-soluble benzene metabolites in F344/N rats and B6C3F$_1$ mice during and following ^{3}H-benzene inhalation. Toxicologist 7:232.

Sabourin, P. J., W. E. Bechtold, L. S. Birnbaum, G. Lucier, and R. F. Henderson. 1988. Differences in the metabolism and disposition of inhaled [3H]benzene by F344/N rats and B6C3F1 mice. Toxicol. Appl. Pharmacol. 94:128–140.

Sackner, M. A., A. Wanner, and J. Landa. 1972. Applications of bronchofiberoscopy. Chest 62(Suppl.):S70–S78.

Sackner, M. A., M. J. Rosen, and A. Wanner. 1973. Estimation of tracheal mucous velocity by bronchofiberscopy. J. Appl. Physiol. 34:495–499.

Sackner, M. A., D. Ford, R. Fernandez, J. Cipley, D. Perez, M. Kwoka, M. Reinhart, E. D. Michaelson, R. Schreck, and A. Wanner. 1978. Effects of sulfuric acid aerosol on cardiopulmonary function of dogs, sheep, and humans. Am. Rev. Respir. Dis. 118:497–510.

Said, S. I. 1982. Pulmonary metabolism of prostaglandins and vasoactive peptides. Annu. Rev. Physiol. 44:257–268.

Saksella, K., J. Bergh, V. P. Lehto, K. Nilsson, and K. Alitalo. 1985. Amplification of the c-myc oncogene in a subpopulation of human small cell lung cancer. Cancer Res. 45:1823–1827.

Saltini, C., A. J. Hance, V. J. Ferrans, F. Basset, P. B. Bitterman, and R. G. Crystal. 1984. Accurate quantification of cells recovered by bronchoalveolar lavage. Am. Rev. Respir. Dis. 130:650–658.

Saltini, C., J. R. Spurzem, J. J. Lee, P. Pinkston, and R. G. Crystal. 1986. Spontaneous release of interleukin 2 by lung T-lymphocytes in active pulmonary sarcoidosis is primarily from the Leu3$^+$DR$^+$ T cell subset. J. Clin. Invest. 77:1962–1970.

Samet, J. M. 1978. A historical and epidemiologic perspective on respiratory symptoms questionnaires. Am. J. Epidemiol. 108:435–446.

Samet, J. M., I. B. Tager, and F. E. Speizer. 1983. The relationship between respiratory illness in childhood and chronic air-flow obstruction in adulthood. Am. Rev. Respir. Dis. 127:508–523.

Sanchis, J., M. Dolovich, R. Chalmers, and M. Newhouse. 1972. Quantitation of regional aerosol clearance in the normal human lung. J. Appl. Physiol. 33:757–762.

Santa Cruz, R., J. Landa, J. Hirsch, and M. A. Sackner. 1974. Tracheal mucous velocity in normal man and patients with obstructive lung disease: Effects of terbutaline. Am. Rev. Respir. Dis. 109:458–463.

Santella, R. M., F. Gasparo, and L. L. Hsieh. 1987. Quantitation of carcinogen-DNA adducts with monoclonal antibodies. Prog. Exp. Tumor Res. 31:63–75.

Santos, E., D. Martin-Zanca, E. P. Reddy, M. A. Pierotti, G. Della Porta, and M. Barbacid. 1984. Malignant activation of a K-ras oncogene in lung carcinoma but not in normal tissue of the same patient. Science 223:661–664.

Sasame, H. A., and M. R. Boyd. 1979. Superoxide and hydrogen peroxide production and NADPH oxidation stimulated by nitrofurantoin in lung microsomes: Possible implications for toxicity. Life Sci. 24:1091–1096.

Schecter, A., T. Tiernan, F. Schaffner, M. Taylor, G. Gitlitz, G. F. VanNess, J. H. Garrett, and D. J. Wagel. 1985. Patient fat biopsies for chemical analysis and liver biopsies for ultrastructural characterization after exposure to polychlorinated dioxins, furans, and PCBs. Environ. Health Perspect. 60:241-254.

Schleimer, R. P., D. W. MacGlashan, Jr., S. P. Peters, N. Naclerio, D. Proud, N. F. Adkinson, Jr., and L. M. Lichtenstein. 1984. Inflammatory mediators and mechanisms of release from purified human basophils and mast cells. J. Allergy Clin. Immunol. 74:473–481.

Schlesinger, R. B. 1989. Deposition and clearance of inhaled particles, pp. 163–192. In R. O. McClellan and R. F. Henderson, Eds. Concepts in Inhalation Toxicology. Washington, D.C.: Hemisphere.

Schlesinger, R. B., and K. E. Driscoll. 1988. Respiratory tract defense mechanisms and their interaction with air pollutants, pp. 37–72. In M. A. Hollinger, Ed. Focus on Pulmonary Pharmacology and Toxicology. Boca Raton, Fla.: CRC Press.

Schlesinger, R. B., M. Lippmann, and R. E. Albert. 1978. Effects of short-term exposures to sulfuric acid and ammonium sulfate aerosols upon bronchial airway function in the donkey. Am. Ind. Hyg. Assoc. J. 39:275–286.

Schlesinger, R. B., M. Halpern, R. E. Albert, and M. Lippmann. 1979. Effect of chronic inhalation of sulfuric acid mist upon mucociliary clearance from the lungs of donkeys. J. Environ. Pathol. Toxicol. 2:1351–1367.

Schlesinger, R. B., M. Halpern, and M. Lippmann. 1982. Long-term clearance of inhaled magnetite and polystyrene latex from the lung: A comparison. Health Phys. 42:68–73.

Schlesinger, R. B., B. Naumann, and L. C. Chen. 1983. Physiological and histological alterations in the bronchial mucociliary clearance system of rabbits following intermittent oral or nasal inhalation of sulfuric acid mist. J. Toxicol. Environ. Health 12:441–465.

Schlesinger, R. B., T. A. Vollmuth, B. D. Naumann, and K. E. Driscoll. 1986. Measurement of particle clearance from the alveolar region of the rabbit respiratory tract. Fundam. Appl. Toxicol. 7:256–263.

Schmidt, J. A., S. B. Mizel, D. Cohen, and I. Green. 1982. Interleukin 1, a potential regulator of fibroblast proliferation. J. Immunol. 128:2177–2182.

Schober, O., G. J. Meyer, C. Bossaller, P. Lobenhoffer, B. Knoop, S. Muller, H. Creutzig, J. Sturm, P. Lichtlen, and H. Hundeshagen. 1983. Quantitative measurements of regional extravascular lung water in dogs using positron emission tomography. ROEFO 139:117–126. (In German.)

Schulman, E. S. 1986. The role of mast cell derived mediators in airway hyper-responsiveness. Chest 90:578–583.

Schulman E. S., and C. L. Anderson. 1985. Lung hypersensitivity responses: Dissociation of human lung mast cell leukotriene C_4 (LTC_4) from histamine release. Prog. Clin. Biol. Res. 199:209–220.

Schulman, E., M. Liu, D. Proud, D. Macglashan, Jr., L. Lichtenstein, and M. Plaut. 1985. Human lung macrophages induce histamine release from basophils and mast cells. Am. Rev. Respir. Dis. 131:230–235.

Schuster, D. P., M. A. Mintun, M. A. Green, and M. M. Ter-Pogossian. 1985. Regional lung water and hematocrit determined by positron emission tomography. J. Appl. Physiol. 59:860–868.

Schuster, D. P., G. F. Marklin, and M. A. Mintun. 1986. Regional changes in extravascular lung water detected by positron emission tomography. J. Appl. Physiol. 60:1170–1178.

Schuyler, M. R., T. P. Thigpen, and J. E. Salvaggio. 1978. Local pulmonary immunity in pigeon breeder's disease: A case study. Ann. Intern. Med. 88:355–358.

Schwartz, L. B., and K. F. Austen. 1984. Structure and function of the chemical mediators of mast cells. Prog. Allergy 34:271–321.

Schwartz, L. B., M. S. Kawahara, T. E. Hugli, D. Vik, D. T. Fearon, and K. F. Austen. 1983. Generation of C3a anaphylatoxin from human C3 by human mast cell tryptase. J. Immunol. 130:1891–1895.

Seaton, A. 1983. Coal and the lung. Thorax 38:241–243.

Sega, G. A., R. P. Valdivia Alcota, C. P. Tanconge, and P. A. Brimer. In press. Acrylamide binding to the DNA and protamine of spermigenic stages in the mouse and its relationship to genetic damage. Mutat Res.

Segerback, D. 1983. Alkylation of DNA and hemoglobin in the mouse following exposure to ethene and ethene oxide. Chem. Biol. Interact. 45:139–151.

Selikoff, I. J., and E. C. Hammond. 1978. Asbestos-associated disease in United States shipyards. CA 28:87–99.

Seltzer, J., B. G. Bigby, M. Stulbarg, M. J. Holtzman, J. A. Nadel, I. F. Ueki, G. D. Leikauf, E. J. Goetzl, and H. A. Boushey. 1986. O_3-induced change in bronchial reactivity to methacholine and airway inflammation in humans. J. Appl. Physiol. 60:1321–1326.

Semenzato, G., A. Pezzutto, M. Chilosi, and G. Pizzolo. 1982. Redistribution of T lymphocytes in the lymph nodes of patients with sarcoidosis. N. Engl. J. Med. 306:48–49.

Sharma, R. P., Ed. 1981. Immunologic Considerations in Toxicology. Boca Raton, Fla.: CRC Press. 2 vols.

Sheppard, D. 1986a. Mechanisms of bronchoconstriction

from nonimmunologic environmental stimuli. Chest 90:584–587.

Sheppard, D. 1986b. Significance of airway hyperresponsiveness to occupational and environmental lung disorders. Semin. Respir. Med. 7:241–248.

Sheppard, D., W. S. Wong, C. F. Uehara, J. A. Nadel, and H. A. Boushey. 1980. Lower threshold and greater bronchomotor responsiveness of asthmatic subjects to sulfur dioxide. Am. Rev. Respir. Dis. 122:873–878.

Shirai, F., S. Kudoh, A. Shibuya, K. Sada, and R. Mikami. 1981. Crackles in asbestos workers: Auscultation and lung sound analysis. Br. J. Dis. Chest 75:386–396.

Shu, H., R. E. Talcott, S. A. Rice, and E. T. Wei. 1979. Lipid peroxidation and paraquat toxicity. Biochem. Pharmacol. 28:327–331.

Shugart, L., and R. Matsunami. 1985. Adduct formation in hemoglobin of the newborn mouse exposed in utero to benzo[a]pyrene. Toxicology 37:241–245.

Sibbald, W. J., F. J. Warshawski, A. K. Short, J. Harris, M. S. Lefcoe, and R. L. Holliday. 1983. Clinical studies of measuring extravascular lung water by the thermal dye technique in critically ill patients. Chest 83:725–731.

Sietsema, K., R. M. Effros, S. T. Siu, and G. R. Mason. 1986. Solute concentrations of bronchoalveolar fluid. Am. Rev. Respir. Dis. 133:A20. (Abstract.)

Simani, A. S., S. Inoue, and J. C. Hogg. 1974. Penetration of the respiratory epithelium of guinea pigs following exposure to cigarette smoke. Lab. Invest. 31:75–81.

Simenhoff, M. L., J. F. Burke, J. J. Saukkonen, A. Ordinario, and R. Doty. 1977. Biochemical profile of uremic breath. N. Engl. J. Med. 297:132–135.

Sivak, E. D., and H. P. Wiedemann. 1986. Clinical measurement of extravascular lung water. Crit. Care Clinics 2:511–526.

Slauson, D. O. 1982. The mediation of pulmonary inflammatory injury. Adv. Vet. Sci. Comp. Med. 26:99–153.

Small, P., D. Barrett, S. Frenkiel, L. Rochon, C. Cohen, and M. Black. 1985. Measurement of antigen-specific IgE in nasal secretions of patients with perennial rhinitis. Ann. Allergy 55:68–71.

Smits, H. L., and A. M. Jetten. In press. Molecular cloning of gene sequences regulated during squamous differentiation of tracheal epithelial cells and controlled by retinoids. J. Biol. Chem.

Snider, D. E., Jr. 1982. The tuberculin skin test. Am. Rev. Respir. Dis. 125:S108–S118.

Snipes, M. B., B. A. Muggenburg, and D. E. Bice. 1983. Translocation of particles from lung lobes or the peritoneal cavity to regional lymph nodes in beagle dogs. J. Toxicol. Environ. Health 11:703–712.

Snyder, A., L. Skoza, and Y. Kikkawa. 1983. Comparative removal of ascorbic acid and other airway substances by sequential bronchoalveolar lavages. Lung 161:111–121.

Snyderman, R. 1981. Chemotaxis of human and murine mononuclear phagocytes, pp. 535–547. In D. O.

Adams, P. J. Edelson, and H. S. Koren, Eds. Methods for Studying Mononuclear Phagocytes. New York: Academic Press.

Soter, N. A., R. A. Lewis, E. J. Corey, and K. F. Austen. 1983. Local effects of synthetic leukotrienes (LTC$_4$, LTD$_4$, LTE$_4$, and LTB$_4$) in human skin. J. Invest. Dermatol. 80:115–119.

Speizer, F. E., and I. B. Tager. 1979. Epidemiology of chronic mucus hypersecretion and obstructive airways disease. Epidmiol Rev 1:124–142.

Spektor, D. M., M. Lippmann, P. J. Lioy, G. D. Thurston, K. Citak, D. J. James, N. Bock, F. E. Speizer, and C. Hayes. 1988. Effects of ambient ozone on respiratory function in active, normal children. Am. Rev. Respir. Dis. 137:313–320.

Spencer, H. 1977. Pathology of the Lung, Excluding Pulmonary Tuberculosis, 3d ed. New York: Pergamon. 2 vols.

Stahlhofen, W., J. Gebhart, J. Heyder, K. Philipson, and P. Camner. 1981. Intercomparison of regional deposition of aerosol particles in the human respiratory tract and their long-term elimination. Exp. Lung Res. 2:131–139.

Stalcup, S. A., G. M. Turino, and R. B. Mellins. 1982. Critical issues in the use of vasoactive substances to assess lung microvascular injury. Ann. N.Y. Acad. Sci. 384:435–457.

Stanley, P. J., R. Wilson, M. A. Greenstone, I. S. Mackay, and P. J. Cole. 1985. Abnormal nasal mucociliary clearance in patients with rhinitis and its relationship to concommitent chest disease. Br. J. Dis. Chest. 79:77–82.

Starkey, P. M., and A. J. Barrett. 1977. α_2-macroglobulin, a physiological regulator of proteinase activity, pp. 663–696. In A. J. Barrett, Ed. Proteinases in Mammalian Cells and Tissues. Amsterdam: Elsevier North-Holland.

Starr, T. B., and J. E. Gibson. 1984. The importance of delivered dose in quantitative risk estimation: Formaldehyde. Toxicologist 4:30. (Abstract.)

Staub, N. C. 1974. Pulmonary edema. Physiol. Rev. 54:678–811.

Staub, N. C. 1986. Clinical use of lung water measurements. Report of a workshop. Chest 90:588–594.

Stern, R. M., K. Drenck, O. Lyngenbo, H. Dirksen, and S. Groth. 1986. Health, occupational exposure and thoracic magnetic moment of shipyard welders, pp. 403–407. In R. M. Stern, A. Berlin, A. C. Fletcher, J. Jarvisalo, Eds. Health Hazards and Biological Effects of Welding Fumes and Gases. International Congress Series, No. 676. Amsterdam: Excerpta Medica.

Stowers, S. J., P. L. Glover, S. H. Reynolds, L. R. Boone, R. R. Maronpot, and M. W. Anderson. 1987. Activation of the K-*ras* protooncogene in lung tumors from rats and mice chronically exposed to tetranitromethane. Cancer Res. 47:3212–3219.

Strumpf, I. J., M. K. Feld, M. J. Cornelius, B. A. Keogh, and R. G. Crystal. 1981. Safety of fiberoptic bronchoalveolar lavage in evaluation of interstitial lung disease. Chest 80:268–271.

Stutts, M. J., and P. A. Bromberg. 1987. Effects of ozone on airway epithelial permeability and ion transport. Toxicol. Lett. 35:315–319.

Stutts, M. J., R. C. Boucher, P. A. Bromberg, and J. T. Gatzy. 1981. Effects of ammonium and nitrate salts on ion transport across the excised canine trachea. Toxicol. Appl. Pharmacol. 60:91–105.

Stutts, M. J., R. C. Boucher, and J. T. Gatzy. 1982. Effects of cadmium and zinc on canine tracheal bioelectric properties. Toxicol. Appl. Pharmacol. 64:147–154.

Stutts, M. J., J. T. Gatzy, M. R. Knowles, C. U. Cotton, and R. C. Boucher. 1986. Effects of formaldehyde on bronchial ion transport. Toxicol. Appl. Pharmacol. 82:360–367.

Sue, D. Y., A. Oren, J. E. Hansen, and K. Wasserman. 1987. Diffusing capacity for carbon monoxide as a predictor of gas exchange during exercise. N. Engl. J. Med. 316:1301–1306.

Sun, J. D., R. K. Wolff, H. M. Aberman, and R. O. McClellan. 1983. Inhalation of 1-nitropyrene associated with ultrafine insoluble particles or as a pure aerosol: A comparison of deposition and biological fate. Toxicol. Appl. Pharmacol. 69:185–198.

Sun, J. D., R. K. Wolff, G. M. Kanapilly, and R. O. McClellan. 1984. Lung retention and metabolic fate of inhaled benzo(a)pyrene associated with diesel exhaust particles. Toxicol. Appl. Pharmacol. 73:48–59.

Sun, J. D., S. S. Ragsdale, J. M. Benson, and R. F. Henderson. 1985. Effects of the long-term depletion of reduced glutathione in mice administered L-buthionine-S,R-sulfoximine. Fundam. Appl. Toxicol. 5:913–919.

Sutherland, M. W., M. Glass, J. Nelson, Y. Lyen, and H. J. Forman. 1985. Oxygen toxicity: Loss of lung macrophage function without metabolite depletion. J. Free Radic. Biol. Med. 1:209–214.

Swenberg, J. A., F. C. Richardson, L. Tyeryar, F. Deal, and J. Boucheron. 1986. The molecular dosimetry of DNA adducts formed by continuous exposure of rats to alkylating hepatocarcinogens. Prog. Exp. Tumor Res. 31:42–51.

Tannenbaum, S. R., and P. L. Skipper. 1984. Biological aspects to the evaluation of risk: Dosimetry of carcinogens in man. Fundam. Appl. Toxicol. 4:S367–S373.

Taplin, G. V., D. P. Tashkin, S. K. Chopra, O. E. Anselmi, D. Elam, B. Calvarese, A. Coulson, R. Detels, and S. N. Rokaw. 1977. Early detection of chronic obstructive pulmonary disease using radionuclide lung-imaging procedures. Chest 71:567–575.

Task Group on Lung Dynamics. 1966. Deposition and retention models for internal dosimetry of the human respiratory tract. Health Phys. 12:173–207.

Tate, R. M., and J. E. Repine. 1983. Neutrophils and the adult respiratory distress syndrome. Am. Rev. Respir. Dis. 128:552–550.

Taylor, S. M., H. Downes, C. A. Hirshman, J. E. Peters, and D. Leon. 1983. Pulmonary uptake of mannitol as an index of changes in lung epithelial permeability. J. Appl. Physiol. 55:614–618.

Teculescu, D. B., Q. T. Pham, and B. Hannhart. 1986. Tests of small airway dysfunction: Their correlation with the "conventional" lung function tests. Eur. J. Respir. Dis. 69:175–187.

Teisinger, J., and V. Fiserova-Bergerova. 1955. Valeur comparée de la détermination des sulfates et du phénol contenus dans l'urine pour l'évaluation de la concentration du benzène dans l'air. [Comparative value of determination of urinary sulfates and phenol for evaluation of atmospheric benzene.] Arch. Mal. Profess. 16:221–232.

Thomas, P. D., and G. W. Hunninghake. 1987. Current concepts of the pathogenesis of sarcoidosis. Am. Rev. Respir. Dis. 135:747–760.

Thomas, P. S. 1983. Hybridization of denatured RNA transferred or dotted to nitrocellulose paper. Methods Enzymol. 100:255–266.

Thomas, Y., R. Huchet, and D. Granjon. 1981. Histamine-induced suppressor cells of lymphocyte mitogenic response. Cell. Immunol. 59:268–275.

Thomson, M. L., and D. Pavia. 1974. Particle penetration and clearance in the human lung: Results in healthy subjects and subjects with chronic bronchitis. Arch. Environ. Health 29:214–219.

Thomson, M. L., and M. D. Short. 1969. Mucociliary function in health, chronic obstructive airway disease, and asbestosis. J. Appl. Physiol. 26:535–539.

Thrall, R. S., J. R. McCormick, R. M. Jack, R. A. McReynolds, and P. A. Ward. 1979. Bleomycin-induced pulmonary fibrosis in the rat: Inhibition by indomethacin. Am. J. Pathol. 95:117–130.

Tingley, D. T., M. Manning, L. C. Grothaus, and W. F. Burns. 1979. The influence of light and temperature on isoprene emission rates from live oak. Physiol. Plant. 47:112–118.

Todorovich, L., D. Johnson, P. Ranger, F. Keeley, and M. Rabinovitch. 1986. Contribution of elastin and collagen to the pathogenesis of monocrotaline induced pulmonary hypertension. Fed. Proc. 45:696. (Abstract.)

Toews, G. B., W. C. Vial, M. M. Dunn, P. Guzzetta, G. Nunez, P. Stastny, and M. F. Lipscomb. 1984. The accessory cell function of human alveolar macrophages in specific T cell proliferation. J. Immunol. 132:181–186.

Togias, A. G., R. M. Naclerio, S. P. Peters, I. Nimmagadda, D. Proud, A. Kagey-Sobotka, N. F. Adkinson, Jr., P. S. Norman, and L. M. Lichtenstein. 1986. Local generation of sulfidopeptide leukotrienes upon nasal provocation with cold dry air. Am. Rev. Respir. Dis. 133:1133–1137.

Toigo, A., J. J. Imarisio, H. Murmall, and M. N. Lepper. 1963. Clearance of large carbon particles from the human tracheobronchial tree. Am. Rev. Respir. Dis. 87:487–492.

Tomioka M., S. Ida, Y. Shinoh, T. Ishihara, and T. Taki-shima. 1984. Mast cells in bronchoalveolar lumen of patients with bronchial asthma. Am. Rev. Respir. Dis. 129:1000–1005.

Toomes, H., I. Vogt-Moykopf, W. D. Heller, and H. Ostertag. 1981. Measurement of mucociliary clearance in smokers and nonsmokers using a bronchoscopic video-technical method. Lung 159:27–34.

Tornqvist, M., S. Osterman-Golkar, A. Kautiainen, S. Jensen, P. B. Farmer, and L. Ehrenberg. 1986. Tissue doses of ethylene oxide in cigarette smokers determined from adduct levels in hemoglobin. Carcinogenesis 7:1519–1521.

U.S. Environmental Protection Agency. 1984. Air Quality Criteria for Ozone and Other Photochemical Oxidants, Vol. IV. Washington, D.C.: U.S. Environmental Protection Agency.

Unanue, E. R., D. I. Beller, C. Y. Lu, and P. M. Allen. 1984. Antigen presentation: Comments on its regulation and mechanism. J. Immunol. 132:1–5.

Utell, M. J., P. E. Morrow, and R. W. Hyde. 1984. Airway reactivity to sulfate and sulfuric acid aerosols in normal and asthmatic subjects. J. Air Pollut. Control Assoc. 34:931–935.

Utell, M. J., P. E. Morrow, M. A. Bauer, G. Oberdörster, D. A. Weber, and R. W. Hyde. 1985. Epithelial permeability following ozone inhalation in healthy subjects. Am. Rev. Respir. Dis. 131:A171. (Abstract.)

Valberg, P. A., and J. D. Brain. 1979. Generation and use of three types of iron-oxide aerosol. Am. Rev. Respir. Dis. 120:1013–1024.

Valenti, S., A. Scordamaglia, P. Crimi, and C. Mereu. 1982. Bronchoalveolar lavage and transbronchial lung biopsy in sarcoidosis and extrinsic allergic alveolitis. Eur. J. Respir. Dis. 63:564–569.

Van oud Alblas, A. B., and R. van Furth. 1979. Origin, kinetics and characteristics of pulmonary macrophages in the normal steady state. J. Exp. Med. 149:1504–1518.

Van Ree, J. H. L., and H. A. E. Van Dishoeck. 1962. Some investigations on nasal ciliary activity. Pract. Otorhinolaryngol. 24:383–390.

Vaughan, T. R., Jr., L. F. Jennelle, and T. R. Lewis. 1969. Long-term exposure to low levels of air pollutants. Effects on pulmonary function in the beagle. Arch. Environ. Health 19:45–50.

Venet, A., A. J. Hance, C. Saltini, B. W. S. Robinson, and R. G. Crystal. 1985. Enhanced alveolar macrophage-mediated antigen-induced T-lymphocyte proliferation in sarcoidosis. J. Clin. Invest. 75:293–301.

Wagenvoort, C. A., and N. Wagenvoort. 1977. Pathology of Pulmonary Hypertension. New York: John Wiley & Sons. 345 pp.

Wagner, J. C. 1965. The sequelae of exposure to asbestos dust. Ann. N.Y. Acad. Sci. 132:691–695.

Wallace, L. A. 1987. The Total Exposure Assessment Methodology (TEAM) Study: Summary and Analysis: Volume I. EPA 600/6-87/002a. Washington, D. C.: U.S. Environmental Protection Agency, Office of Acid Depositon, Environmental Monitoring and Quality Assurance.

Wanner, A. 1977. Clinical aspects of mucociliary transport. Am. Rev. Respir. Dis. 116:73–125.

Wanner, A. 1980. Pulmonary defense mechanisms: Mucociliary clearance, pp. 325–356. In D. H. Simmons, Ed. Current Pulmonology, Vol. 2. Boston: Houghton Mifflin.

Ware, J. H., D. W. Dockery, A. Spiro, III, F. E. Speizer, and B. G. Ferris, Jr. 1984. Passive smoking, gas cooking, and respiratory health of children living in six cities. Am. Rev. Respir. Dis. 129:366–74.

Warheit, D. B., L. Y. Chang, L. H. Hill, G. E. R. Hook, J. D. Crapo, and A. R. Brody. 1984. Pulmonary macrophage accumulation and asbestos-induced lesions at sites of fiber deposition. Am. Rev. Respir. Dis. 129:301–310.

Warheit, D. B., G. George, L. H. Hill, R. Snyderman, and A. R. Brody. 1985. Inhaled asbestos activates a complement-dependent chemoattractant for microphages. Lab. Invest. 52:505–514.

Warheit, D. B., L. H. Hill, G. George, and A. R. Brody. 1986. Time course of chemotactic factor generation and the corresponding macrophage response to asbestos inhalation. Am. Rev. Respir. Dis. 134:128–133.

Warheit, D. B., L. H. Overby, G. George, and A. R. Brody. 1988. Pulmonary macrophages are attracted to inhaled particles through complement activation. Exp. Lung Res. 14:51–66.

Warner, J., M. M. Pienkowski, M. Plaut, P. S. Norman, and L. M. Lichtenstein. 1986. Identification of histamine releasing factor(s) in the late phase of cutaneous IgE-mediated reactions. J. Immunol. 136:2583–2587.

Wasserman, K. 1985. An integrative approach to the non-invasive assessment of cardiovascular and respiratory function during exercise. Workshop on Physical Fitness and Activity Assessment in NCHS General Population Surveys, National Center for Health Statistics, June.

Wasserman, K., B. J. Whipp, and J. A. Davis. 1981. Respiratory physiology of exercise: Metabolism, gas exchange, and ventilatory control. Int. Rev. Physiol. 23:149–211.

Wasserman, K., J. E. Hansen, D. Y. Sue, and B. J. Whipp. 1987. Principles of Exercise Testing and Interpretation. Philadelphia: Lea and Febiger.

Weibel, E. R. 1979. Stereological Methods: Practical Methods for Biological Morphometry, Vol. 1. New York: Academic Press.

Weinberger, S. E., J. A. Kelman, N. A. Elson, R. C. Young, Jr., H. Y. Reynolds, J. D. Fulmer, and R. G. Crystal. 1978. Bronchoalveolar lavage in interstitial lung disease. Ann. Intern. Med. 89:459–466.

Weiss, J. W., J. M. Drazen, N. Coles, E. R. McFadden, Jr., P. F. Weller, E. J. Corey, R. A. Lewis, and K. F. Austen. 1982. Bronchoconstrictor effects of leukotriene C in humans. Science 216:196–198.

Weiss, S. T., I. B. Tager, F. E. Speizer, and B. Rosner. 1980. Persistent wheeze: Its relation to respiratory illness, cigarette smoking, and level of pulmonary function in a population sample of children. Am. Rev. Respir. Dis. 122:697—707.

Weiss, S. T., I. B. Tager, J. W. Weiss, A. Munoz, F. E. Speizer, and R. H. Ingram. 1984. Airways responsiveness in a population sample of adults and children. Am. Rev. Respir. Dis. 129:898—902.

Weissler, J. C., C. R. Lyons, M. F. Lipscomb, and G. B. Toews. 1986. Human pulmonary macrophages: Functional comparison of cells obtained from whole lung and by bronchoalveolar lavage. Am. Rev. Respir. Dis. 133:473—477.

Weller, P. F., C. W. Lee, D. W. Foster, E. J. Corey, K. F. Austin, and R. A. Lewis. 1983. Generation and metabolism of 5-lipoxygenase pathway leukotrienes by human eosinophils: Predominant production of leukotriene C4. Proc. Natl. Acad. Sci. USA 80:7626—7630.

Wewers, M. D., S. I. Rennard, A. J. Hance, P. B. Bitterman, and R. G. Crystal. 1984. Normal human alveolar macrophages obtained by bronchoalveolar lavage have a limited capacity to release interleukin-1. J. Clin. Invest. 74:2208—2218.

Wexter, H. R., R. L. Nicholson, F. S. Prato, L. S. Carey, S. Vinitski, and L. Reese. 1985. Quantitation of lung water by nuclear magnetic resonance imaging. A preliminary study. Invest. Radiol. 20:583—590.

White, R. T., D. Damm, J. Miller, K. Spratt, J. Schilling, S. Hawgood, B. Benson, and B. Cordell. 1985. Isolation and characterization of the human pulmonary surfactant apoprotein gene. Nature 317:361—363.

Williams, W. J., and W. R. Williams. 1983. Value of beryllium lymphocyte transformation tests in chronic beryllium disease and in potentially exposed workers. Thorax 38:41—44.

Williams, W. R., and W. J. Williams. 1982. Development of beryllium lymphocyte transformation tests in chronic beryllium disease. Int. Arch. Allergy Appl. Immunol. 67:175—180.

Williamson, S. J., and L. Kaufman. 1981. Biomagnetism topical review. J. Magn. Magn. 22:129—201.

Willis, R. J., and C. C. Kratzing. 1976. Extracellular ascorbic acid in lung. Biochim. Biophys. Acta 444:108—111.

Wilson, J. T., P. F. Milner, M. E. Summer, F. S. Nallaseth, H. E. Fadel, R. H. Reindollar, P. G. McDonough, and L. B. Wilson. 1982. Use of restriction endonucleases for mapping the allele for betas-globin. Proc. Natl. Acad. Sci. USA 79:3628—3631.

Wilson, M. R., H. R. Gaumer, and J. E. Salvaggio. 1977. Activation of the alternate complement pathway and generation of chemotactic factors by asbestos. J. Allergy Clin. Immunol. 60:218—222.

Witz, G., M. A. Amoruso, and B. D. Goldstein. 1983. Effect of ozone on alveolar macrophage function: Membrane dynamic properties, pp. 263—272. In S. D. Lee, M. G. Mustafa, and M. A. Mehlman, Eds. Proceedings of the Symposium on the Biomedical Effects of Ozone and Related Photochemical Oxidants,

March 1982, Pinehurst, N.C. Advances in Modern Environmental Toxicology, Vol. 5. Princeton, N.J.: Princeton Scientific Publishers.

Wogan, G. N. 1988. Methods for detecting DNA damaging agents in humans. Summary: Methods. IARC Sci. Publ. 89:9—12.

Wogan, G. N., and N. J. Gorelick. 1985. Chemical and biochemical dosimetry of exposure to genotoxic chemicals. Environ. Health Perspect. 62:5—18.

Wolff, M. S., J. Thornton, A. Fischbein, R. Lilis, and I. J. Selikoff. 1982. Disposition of polychlorinated biphenyl congeners in occupationally exposed persons. Toxicol. Appl. Pharmacol. 62:294—306.

Wolff, R. K. 1986. Effects of airborne pollutants on mucociliary clearance. Environ. Health Perspect. 66:223—237.

Wolff, R. K., B. A. Muggenberg, and S. A. Silbaugh. 1981. Effect of 0.3 and 0.9 micron sulfuric acid aerosols on tracheal mucous clearance in beagle dogs. Am. Rev. Respir. Dis. 123:291—294.

Wolff, R. K., B. A. Muggenburg, S. A. Silbaugh, R. L. Carpenter, J. D. Rowatt, and J. O. Hill. 1982. Lung Clearance Mechanisms Following Inhalation of Acid Sulfates. Report LMF-99. Albuquerque, N.M.: Inhalation Toxicology Research Institute, Lovelace Biomedical and Environmental Research Institute. Available from NTIS.

Wolff, R. K., R. F. Henderson, A. F. Eidson, J. A. Pickrell, S. J. Rothenberg, and F. F. Hahn. 1988. Toxicity of gallium oxide particles following a 4-week inhalation exposure. J. Appl. Toxicol. 8:191—199.

Wolff, R. K., J. A. Bond, J. D. Sun, R. F. Henderson, J. R. Harkema, W. C. Griffith, J. L. Mauderly, and R. O. McClellan. 1989. Effects of adsorption of benzo[a]pyrene onto carbon black particles on levels of DNA adducts in lungs of rats exposed by inhalation. Toxicol. Appl. Pharmacol. 97:289—299.

Wollmer, P., C. G. Rhodes, and J. M. B. Hughes. 1984. Regional extravascular density and fractional blood volume of the lung in interstitial disease. Thorax 39:286—293.

Woo, S. L. C., A. S. Lidsky, F. Guttler, T. Chandra, and K. J. H. Robson. 1983. Cloned human phenylalanine hydroxylase gene allows prenatal diagnosis and carrier detection of classical phenylketonuria. Nature 306:151—155.

Wood, P. B., E. Nagy, and F. G. Pearson. 1973. Measurement of mucociliary clearance from the lower respiratory tract of normal dogs. Can. Anaesth. Soc. J. 20:192—206.

Woolcock, A. J., N. J. Vincent, and P. T. Macklem. 1969. Frequency dependence of compliance as a test for obstruction in the small airways. J. Clin. Invest. 48:1097—1106.

Wooten, F. T., W. W. Warring, M. J. Wegmann, W. F. Anderson, and J. D. Conley. 1978. Methods for respiratory sound analysis. Med. Instrum. 12:254—257.

Yanagisawa, Y., H. Nishimura, H. Matsuki, F. Osaka, and H. Kasuga. 1986. Personal exposure and health effect relationship for NO_2 with urinary hydroxyproline to creatinine ratio as indicator. Arch. Env. Health 41:41—48.

Yeates, D. B., M. Aspin, H. Levison, M. T. Jones, and A. C. Bryan. 1975. Mucociliary tracheal transport rates in man. J. Appl. Physiol. 39:487—495.

Young, K. R., and H. Y. Reynolds. 1984. Bronchoalveolar washings: Proteins and cells from normal lungs, pp. 157—173. In J. Bienenstock, Ed. Immunology of the Lung and Upper Respiratory Tract. New York: McGraw-Hill.

Zarbl, H., S. Sukumar, A. V. Arthur, D. Martin-Zanca, and M. Barbacid. 1985. Direct mutagenesis of the HA-ras-1 oncogenes by N-nitroso-N-methylurea during initiation of mammary carcinogenesis in rats. Nature 315:382—385.

Zarins, L. P., and J. L. Clausen. 1982. Body plethysmography, pp. 141—153. In J. L. Clausen, Ed. Pulmonary Function Testing Guidelines and Controversies: Equipment, Methods, and Normal Values. New York: Academic Press.

Zavala, D. C. 1978. Flexible Fiberoptic Bronchoscopy. A Training Handbook. Iowa City: University of Iowa. 165 pp.

Zeiss, C. R., P. Wolkonsky, R. Chacon, P. A. Tuntland, D. Levitz, J. J. Prunzansky, and R. Patterson. 1983. Syndromes in workers exposed to trimetallic anhydride: A longitudinal clinical and immunologic study. Ann. Intern. Med. 98:8—12.

Zlatkis, A., R. S. Brazell, and C. F. Poole. 1981. The role of organic volatile profiles in clinical diagnosis. Clin. Chem. 27:789—797.

Biographies

ROGENE F. HENDERSON, Chairman, Subcommittee on Pulmonary Toxicology, is a senior scientist and supervisor of the Chemistry and Biochemical Toxicology Group at the Lovelace Inhalation Toxicology Research Institute in Albuquerque, New Mexico. She has done extensive research on the analysis of bronchoalveolar lavage fluid to evaluate lung injury in animal toxicology studies. She has also headed studies on the disposition and metabolic fate of inhaled vapors to aid in planning and interpretation of long-term carcinogenicity studies in rodents. Dr. Henderson has been a member of the Committee on Toxicology of the Board on Environmental Studies and Toxicology of the National Research Council since 1985.

MARIE AMORUSO is an assistant professor in the Department of Environmental and Community Medicine at University of Medicine and Dentistry of New Jersey-Robert Wood Johnson Medical School and a member of the faculty of the joint graduate program in toxicology, medical school and Rutgers University. She is a toxicologist whose major research interests have been in health effects of air pollutants. Her research has focused on mechanisms of oxidant damage in cell membranes and the effects of the oxidant air pollutants ozone and nitrogen dioxide on alveolar macrophage membrane structure and function. She is also a co-principal investigator in studies on the bioavailability of dioxin and its transpulmonary absorption from lung tissue and is developing new models for risk assessment of pulmonary toxicants.

ARNOLD R. BRODY is senior scientist and head of the Pulmonary Pathology Laboratory, National Institute of Environmental Health Sciences. He received an M.S. from the University of Illinois in 1967 and a Ph.D. from Colorado State University in 1969, and he held a post doctoral position at Ohio State University in 1969-1972. Dr. Brody is an adjunct professor in the Department of Pathology, graduate school faculty, and curriculum in toxicology of Duke University College of Medicine. He also holds an adjunct appointment in the graduate school faculty and curriculum in toxicology of the University of North Carolina, Chapel Hill.

EDWARD D. CRANDALL is chief of the Division of Pulmonary and Critical Care Medicine and Bruce Webster Professor of Internal Medicine at the New York Hospital, Cornell Medical Center, New York, New York. He won a research career development award from NIH in 1975.

Dr. Crandall is a member of the American College of Physicians, the American Thoracic Society, the American Society for Clinical Investigation, the American Physiological Society, and the American Institute of Chemical Engineering. His research interests are in lung epithelial transport, pulmonary physiology and disease, pulmonary gas exhange, and cell membrane structure and function.

JAMES D. CRAPO is a professor of medicine and associate professor of pathology at Duke University Medical Center. He received his M.D. degree from the University of Rochester in 1971. He served an internship in internal medicine at Harvard General Hospital in Torrence, California, and a residency in internal medicine followed by a fellowship in pulmonary medicine at Duke University Medical Center. He served as a staff associate at the National Institute of Environmental Health Sciences from 1972 to 1975. In 1976, Dr. Crapo joined the faculty at Duke University Medical Center, where he has served as chief of the Division of Allergy, Critical Care and Respiratory Medicine since 1979. His research interests involve the use of biochemical, immunocytochemical, and ultrastructural techniques to evaluate the effects of free radicals in causing acute lung injury and to determine the effects of exposure to common environmental pollutants on pulmonary structure and function.

RONALD P. DANIELE is professor of medicine and pathology at the Hospital of the University of Pennsylvania. His principal research interests are in the immune inflammatory mechanisms of the lung and their role in mediating pulmonary interstitial lung diseases, in particular acute and chronic lung injury produced by inhaled inorganic agents, such as silica and beryllium. He has studied the effects of those agents in both humans and experimental models. He was one of the first to use bronchoalveolar lavage as an investigative tool for the study of interstitial lung diseases. Dr. Daniele is director of the Pulmonary Immunology Laboratory at the University of Pennsylvania Medical School. He is on the editorial board of *Experimental Lung Research* and is associate editor of the *Journal of Applied Physiology*. In addition to serving on advisory committees of the National Research Council and the Environmental Protection Agency, he is a member of the Pulmonary Diseases Advisory Committee of the National Heart, Lung, and Blood Institute.

CARL FRANZBLAU is a professor and chairman of the Department of Biochemistry at Boston University School of Medicine. He received his Ph.D. in biochemistry from the Albert Einstein College of Medicine in 1962. His research interests are focused on connective-tissue biochemistry and its relation to pulmonary disease.

ALLEN G. HARMSEN is an assistant scientist at the Trudeau Institute in Saranac Lake, New York. He received his Ph.D. in immunobiology from Iowa State University in 1979. He was a postdoctoral fellow at Vermont Lung Center in 1979-1981, assistant professor of biology at New Mexico Institute of Mining and Technology in 1981-1984, and staff scientist at Lovelace Inhalation Toxicology Research Institute in 1984-1986. His current research interests include lung defense mechanisms in the compromised host. Dr. Harmsen belongs to the Reticuloendothelial Society and the American Society for Microbiology.

GARY W. HUNNINGHAKE is a professor of medicine and director of the Division of Pulmonary Medicine at the University of Iowa College of Medicine. His basic research interests are in the molecular mechanisms of lung injury in patients with asthma and interstitial lung disease. He has also pioneered the use of bronchoalveolar lavage in clinical and research studies in patients with those disorders. He received his M.D. degree and his training in internal medicine from the University of Kansas. After leaving Kansas, he spent 3 years as a clinical associate in the National Institute of Allergy and Infectious Diseases and then joined the National Heart, Lung, and Blood Institute as a senior investigator before joining the faculty at the University of Iowa. He has served as president of the American Federation for Clinical Research and is a member of the American Society for Clinical Investigation and the Association of American Physicians. He has also served as a member of the administrative board of the Council of Academic Societies of the Association of American Medical Societies and has served on several NIH study sections.

PHILIP J. LANDRIGAN is a professor of community medicine and director of the Division of Environmental and Occupational Medicine of the Mount Sinai School of Medicine, New York, New York. He obtained his medical degree from the Harvard Medical School in 1967. He completed an internship at Cleveland Metropolitan General Hospital in 1967-1968 and he completed a residency in pediatrics at the Children's Hospital Medical Center, Boston, Massachusetts, in 1968-1970. He is board-certified in pediatrics. From 1970 through 1985, Dr. Landrigan served as a commissioned officer in the U. S. Public Health Service. In 1985, he assumed his present post at the Mount Sinai School of Medicine, where he is responsible for directing a research program in environmental and occupational medicine, for the training of residents, and for the teaching of medical students and he is a professor of pediatrics.

JOE L. MAUDERLY is president of Lovelace Biomedical and Environmental Research Institute and director of the Inhalation Toxicology Research Institute, Albuquerque, New Mexico. He received his doctorate in veterinary medicine from Kansas State University in 1967 and served in the U.S. Air Force at the U.S. Army Natick Laboratories, Natick, Massachusetts, during 1967-1969. He joined the Inhalation Toxicology Research Institute in 1969 as a respiratory physiologist. His primary research interests have been in the effects of inhaled toxic materials on respiratory function and lung structure. He has worked to adapt human respiratory function measurements for use with animals and has studied the usefulness of animal respiratory function data for predicting potential health effects in humans. He has focused primarily on cause-effect relationships between structural and functional changes in the lung. His interests also encompass pulmonary carcinogenesis from inhaled complex materials, animal models of human lung diseases, clinical applications of lung function measurements, influences of age and sex on lung structure and function, and research management.

ROBERT A. ROTH is a professor in the Department of Pharmacology and Toxicology at Michigan State University. He received his Ph.D. in 1975 from Johns Hopkins University. He was a toxicological screening specialist with the U.S. Army Environmental Hygiene Agency in 1969-1971 and a research fellow at Yale University in 1975-1977. He has been at Michigan State University since 1977. Dr. Roth is a member of the American Society for Pharmacology and Experimental Therapeutics, a member of the Society of Toxicology, a diplomate of the American Board of Toxicology, and a member of the NIH Toxicology Study Section. He received the NIH merit award for 1988-1998. Dr. Roth's research interests include mechanisms of lung and liver injury from toxic chemicals and physiologic and toxicologic influences on drug metabolism.

RICHARD B. SCHLESINGER is a professor of environmental medicine at New York University School of Medicine, New York, New York, and director of the Systemic Toxicology Program and Laboratory of Pulmonary Biology and Toxicology, Institute of Environmental Medicine, NYU Medical Center. He received his Ph.D. in 1975 from New York University. Dr. Schlesinger was the recipient of the Kenneth Morgareidge Award for contributions to the field of toxicology in 1987 and of a research career development award from the National Institute of Environmental Health Sciences in 1983. Dr. Schlesinger served on the Environmental Protection Agency Science Advisory Board Extrapolation Models Subcommittee and the National Council on Radiation Protection and Measurements Respiratory Tract Modelling Task Group and was a Consultant to EPA for an issue paper on acid aerosols, state-of-the-science report on direct health effects of air pollutants associated with acidic precursor emissions, and the nitrogen oxides criteria document. He is on the editorial boards of *Inhalation Toxicology* and *Fundamental and Applied Toxicology*. Dr. Schlesinger is vice-president-elect of the Inhalation Specialty Section, Society of Toxicology, and is a member of the Technical Committee, Society of Toxicology.

FRANK E. SPEIZER is a professor of medicine at Harvard Medical School, professor of environmental science at Harvard School of Public Health, and co-director of the Channing Laboratory, Department of Medicine, Brigham and Womens Hospital, Harvard Medical School, Boston, Massachussetts. He received his M.D. from Stanford University Medical School in 1960. His research interests include epidemiologic studies of chronic diseases associated with environmental exposure, particularly heart and lung disease and cancer.

MARK I. UTELL is a professor of medicine and toxicology at the University of Rochester School of Medicine and co-director of the pulmonary disease unit. He also serves as co-director of the occupational health program at the medical school. He received his M.D. from Tufts University School of Medicine in 1972 and his training in pulmonary medicine training at the University of Rochester. His research interests have focused on the pulmonary effects of inhaled environmental pollutants. He serves as a member of the Research Committee of the Health Effects Institute and is a member of the Clean Air Scientific Advisory Committee of the Environmental Protection Agency. He has been an ad hoc member of several NIH study sections and EPA review committees.

Index